Ecological Orbits

Ecological Orbits

How Planets Move and Populations Grow

LEV GINZBURG
MARK COLYVAN

OXFORD
UNIVERSITY PRESS

2004

OXFORD
UNIVERSITY PRESS

Oxford   New York
Auckland   Bangkok   Buenos Aires   Cape Town   Chennai
Dar es Salaam   Delhi   Hong Kong   Istanbul   Karachi   Kolkata
Kuala Lumpur   Madrid   Melbourne   Mexico City   Mumbai   Nairobi
São Paulo   Shanghai   Taipei   Tokyo   Toronto

Text Copyright © 2004 by Applied Biomathematics
Artwork by Amy Dunham © 2002 Applied Biomathematics

Published by Oxford University Press, Inc.,
198 Madison Avenue, New York, New York 10016

www.oup.com

Oxford is a registered trademark of Oxford University Press

All rights reserved. No part of this publication
may be reproduced, stored in a retrieval system, or transmitted,
in any form or by any means, electronic, mechanical,
photocopying, recording, or otherwise, without the prior
permission of Oxford University Press.

Library of Congress Cataloging-in-Publication Data
Ginzburg, Lev R.
Ecological orbits: how planets move and populations grow /
Lev Ginzburg, Mark Colyvan.
p.   cm.
Includes bibliographical references (p.).
ISBN 0-19-516816-X
1. Population biology.   2. Ecology.   I. Colyvan, Mark.   II. Title.
QH352.G55   2003
577.8'8—dc21   2003048690

1 3 5 7 9 8 6 4 2
Printed in the United States of America
on acid-free recycled paper

To Tatyana (1946–2000)

Difficulty in imagining how theory can adequately describe nature is not a proof that theory cannot.
—Robert MacArthur (1930–1972)

Neruda: *Le metafore? . . . Quando parli di una cosa, la paragonono ad un' altra. . . .*
Postino: *É semplice! Perchè questo nome é cosí complicato?*

[Neruda: *Metaphors? . . . It's when you speak of one thing, comparing it to another. . . .*
Postino: *That's simple! Why do they use such a complicated name?*]

—Dialogue between the characters Pablo Neruda and the postman in the movie *Il Postino* (1995)

Preface

The main focus of this book is the presentation of the "inertial" view of population growth. This view provides a rather simple model for complex population dynamics, and is achieved at the level of the single species, without invoking species interactions. An important part of our account is the maternal effect. Investment of mothers in the quality of their daughters makes the rate of reproduction of the current generation depend not only on the current environment but also on the environment experienced by the preceding generation.

The inertial view is a significant departure from traditional ecological theory, which has been developing within the Lotka–Volterra framework for close to a century. One way to see this departure is to focus attention away from the growth rate as the sole variable responding to the environment, and toward "acceleration," or the rate of change of the growth rate between consecutive generations. More precisely, our suggestion is that population growth is a second-order dynamic process at the single-species level, and the second-order character is not necessarily the result of species interactions. If the inertial view of population growth proves correct, a great deal of current theory on population growth will need to be rethought and revised.

As will become clear, our inspiration for looking at ecology in the way we do comes from similar moves in physics—in particular, the move from Aristotelian to Newtonian physics. So let us say a few words in defense of our apparent "physics envy." Many biologists and ecologists find deference to physics, as the

Alfred James Lotka (1880–1986)

science to which all other sciences must aspire, somewhat distasteful. Those who find this deference to physics unreasonable are generally concerned by the inappropriateness of the methods of physics to other branches of science. They suggest that biology, for example, would be better served if biologists did *biology* instead of trying to mimic the methods and successes of physics. We agree that the methods of the various sciences are quite different and that, in general, the methods of physics are of little use to biologists. But this does not mean that the various branches of science cannot take inspiration from one another.

PREFACE   ix

Vito Volterra (1860–1940)

The inspiration here is metaphorical, not mechanical. We see an abstract connection between a certain innovation in the development of physics around the time of Galileo and an analogous way to approach population growth. We are not suggesting any mechanistic similarities between the ways in which populations grow and the ways planets move. At the same time, we are not trading in mere metaphors. Although the inspiration for looking at ecology in the way we do comes from metaphorical connections with physics, these connections can, and will, be spelled out via the mathematics employed in the respective theories we present.

Down through the ages, metaphors and analogies have been a common device for generating new and fruitful ideas in science. We hope, in the end, you will agree with us that the predominant analogy of this book—the analogy between inertial population growth and classical mechanics—is, at the very least, interesting. This analogy helps motivate and shed light on a fruitful way of thinking about population growth. Of course, eventually, the inertial theory of population growth needs to stand on its own merits, and we hope we have contributed to that enterprise as well, but at the same time, we feel no shame in revealing the more poetic beginnings of our thinking on this subject.

Appreciations

First, we thank Kluwer Academic Publishers and Blackwell Publishers for permission to reproduce material from our articles (Colyvan and Ginzburg 2003a, 2003b). Material from chapters 1 and 3 originally appeared in *Biology and Philosophy*, and material from chapter 2 originally appeared in *Oikos*. The copyrights for the material in question remain with Kluwer and Blackwell, respectively.

Next, we owe appreciation to Resit Akçakaya, André Coelho-Levy, Amy Dunham, Dan Dykhuizen, Scott Ferson, Doug Futuyma, Alex Ginzburg, Janos Hajagos, Charlie Janson, Stephan Munch, Bertram Murray, Yamina Oomen, Helen Regan, Justin Roman, Marina Shokhor, Larry Slobodkin, Justine Tietjen, and George Williams for their attendance at, and valuable contributions to, a series of seminars in Old Field, New York, in 2001 devoted to the topic of this book.

We are indebted to Peter Abrams, Resit Akçakaya, Alan Berryman, Rudy Boonstra, Leah Ceccarelli, Amy Dunham, Scott Ferson, David Goodstein, Chris Jensen, John True, Troy Tucker, and Peter Turchin for reading earlier drafts of the whole book manuscript, and for their insightful comments, criticisms, and suggestions. We are also indebted to Roger Arditi, John Damuth, Pablo Inchausti, John Kadvani, Elizabeth Milliman, Helen Regan, Larry Slobodkin, and Justine Tietjen for reading and commenting on drafts of various sections of the book.

Many of the ideas in this book have been developed over a number of years, and many conversations and exchanges with various people have helped clarify our thinking on these matters. These people include Evelyn Fox Keller, Alan Hájek, Greg Mikkelson, Kim Sterelny, and Bill Wimsatt.

We are also indebted to Carol Booth for her help with the index, to Amy Dunham for her artwork, to Elizabeth Milliman for data analysis and editorial assistance, to Herb Mummers for editorial and technical assistance, to Edward Beltrami and Patrícia Maragliano for translating and transcribing the dialogue from *Il Postino*, and to John Damuth and Justin Roman for help with the book title.

Finally, Lev Ginzburg would like to acknowledge a huge debt to Tatyana Ginzburg. For 30 years Tatyana argued that it was a mistake to use physical analogies in advancing an ecological theory. As even a casual reader of this book will notice, Lev ignored her advice. But her advice was not in vain. Her challenge to the use of physical analogies resulted in considerable reflection on the particular analogies employed in this book and, more generally, on the nature and role of analogy in science. This, in turn, resulted in a much clearer book than what would have been written without Tatyana's criticisms. This book is dedicated to Tatyana and her well-intentioned advice.

Contents

1 On Earth as It Is in the Heavens 3
    1.1 How Planets Move 4
    1.2 How Populations Grow 6
    1.3 Metaphors and the Language of Science 8
    1.4 Inertial Population Growth 9

2 Does Ecology Have Laws? 11
    2.1 Ecological Allometries 12
    2.2 Kepler's Laws 21
    2.3 What Is a Law of Nature? 26
    2.4 Laws in Ecology 30

3 Equilibrium and Accelerated Death 34
    3.1 Accelerated Death 35
    3.2 Galileo and Falling Bodies 36
    3.3 The Slobodkin Experiment 39
    3.4 Falling Bodies and Dying Populations 42
    3.5 The Meaning of Abundance Equilibrium 43
    3.6 The Damuth Allometry 46
    3.7 A Harder Question 48

4 The Maternal Effect Hypothesis   49

- 4.1 Inertial Growth and the Maternal Effect   50
- 4.2 The Missing Periods   52
- 4.3 The Calder Allometry   57
- 4.4 The Eigenperiod Hypothesis   59
- 4.5 What Can Be Done in the Laboratory   62

5 Predator–Prey Interactions and the Period of Cycling   64

- 5.1 An Alternative Limit Myth   65
- 5.2 Prey-Dependent versus Ratio-Dependent Models   66
- 5.3 The Fallacy of Instantism   70
- 5.4 Why Period Travels Bottom Up   74
- 5.5 Competing Views on Causes and Cyclicity   78

6 Inertial Growth   83

- 6.1 The Implicit Inertial-Growth Model   83
- 6.2 Parametric Specification   90
- 6.3 Malthusian Invariancy   95
- 6.4 What Is and What Is Not Analogous   100

7 Practical Consequences   104

- 7.1 Theoretical and Applied Ecology   104
- 7.2 Managing Inertial Populations   106
- 7.3 Rates of Evolution   111

7.4　Risk Analysis　113

　　7.5　The Moral　114

8　Shadows on the Wall　117

　　8.1　Plato's Cave　118

　　8.2　Evidence and Aesthetics　120

　　8.3　Overfitting　123

　　8.4　A Simplified Picture of Population Ecology　125

Appendix A: Notes and Further Reading　133

Appendix B: Essential Features of the Maternal Effect Model　143

Bibliography　151

Index　161

Ecological Orbits

ONE

ON EARTH AS IT IS IN THE HEAVENS

Populations grow and decline; planets roll relentlessly around the sun. On the face of it, planets and populations have nothing to do with one another. Indeed, they are studied by completely different branches of science. Planets are studied by a branch of physics—*astronomy*; populations of living organisms are studied by a branch of biology—*population ecology*. And these two disciplines have little in common. Any suggestion that one theory describes both planetary motion and population growth is surely misguided.

There are some similarities between planets and populations, however. For a start, the laws of nature are meant to hold everywhere in the universe. We expect planets in far-off galaxies to be governed by the same natural laws as our own Earth. And we expect theories of population growth to hold for whatever strange creatures inhabit those far-off planets, just as these theories hold for rabbits, bacteria, and humans here on Earth. Second, both planets and populations are capable of periodic behavior. Planets trace out elliptical orbits around the sun and (more or less) repeat these orbits, taking (more or less) the same period of time to complete each cycle. The abundance of certain populations can also rise and fall in a periodic way, so that the number of organisms cycles between extreme values over some (more or less) fixed period of time. Of course, the mechanisms in each case are totally different, but at some level there are striking similarities between planets and at least certain types of populations. In fact, mathematically, the differences between orbits and cycles are rather minor. This suggests that we may find developments in one area of science that provide useful hints for new ways to look at another. In particular, we might find useful clues in physics for the appropriate conceptual framework to adopt in population ecology. It is this latter suggestion that we explore in this book.

We should make it absolutely clear, however, that we do not think that populations *literally* grow in the same way that bodies move. Indeed, that doesn't even make sense. What we propose is that an understanding of why certain developments in physics were so successful will help us to make analogous moves in the advancement of population ecology. We use planetary orbits as a central and important metaphor to guide us in thinking about such matters. Our ultimate aim, however, is to outline some of the recent developments in population ecology and to suggest that we may be on the verge of discovering some general principles governing population growth.

1.1 How Planets Move

As we've already mentioned, our central analogy is that of planetary orbits. In particular, we take our lead from certain developments in physics around the 17th century. An extremely important change of perspective was heralded by Galileo and Newton, who saw uniform motion as the default case. (We will call this the *inertial view*.) The received wisdom at the time was the Aristotelian view that rest was the default state of all bodies. So, according to Aristotle, if a body was in motion, there must be a force acting upon it. According to the inertial view, however, a body in motion would remain in motion (because of *inertia*) unless acted upon by a force. On the Aristotelian view, forces bring about velocities, whereas on the inertial view, forces bring about changes in velocities (or accelerations).

Nowadays it is hard to appreciate just how radical was the introduction of the inertial view of Galileo and Newton. The

Galileo–Newton insight was particularly astonishing because their view seemed to fly in the face of almost all of our everyday experience. After all, even on perfectly flat ground, moving balls come to rest unless sustained by some external force. Rest really does seem to be the natural state of things. But now think about some of the things that do not seem to have a tendency to slow down, the planets, for instance. How is it that planets keep orbiting the sun? According to the Aristotelian view, they must be sustained by some force. Moreover, this force must be constantly changing direction so as to keep them orbiting. They need, as it were, "the hand of God."

According to the inertial view, planets do not need any external force to keep them moving, but their constant change in velocity does need some explanation. The point is simply that, although almost all earthly phenomena seemed to support Aristotelian physics, a couple of key examples, such as the orbits of the planets, were enough to cast doubt on the received view. The new inertial view made more sense of certain, somewhat rare, phenomena, but a story was required to explain why the world *appeared* as though it were Aristotelian. The latter, of course, was a story about frictional forces: on Earth, friction is so prevalent that moving bodies need external forces to sustain their motion, not because rest is the natural state, but because bodies are subject to frictional forces and the external forces must be applied to counter these.

There are a couple of valuable, general lessons to be learned from this. The first is that it is very important to establish what the default theory should be—what happens when nothing happens. The default position, however, is not always the obvious choice. Rest seems like a much more intuitive state for bodies with no forces acting upon them. The real default state, somewhat surprisingly, turns out to be uniform motion, of which rest is only one special case. The second lesson is that sometimes it can be extremely fruitful to focus attention on the exceptions and anomalies. After all, truly inertial motion (i.e., frictionless motion) is hard to find. Nevertheless, it is precisely these rare cases, such as planetary motion, that give us the greatest insights into the laws of mechanics. It is an interesting question whether we would have ever discovered Newtonian physics if we never saw other planets—if we lived on a planet like Venus, for instance, with constant, thick cloud cover.

1.2 How Populations Grow

Now let's consider the ways populations grow. Populations whose abundances are cyclic are a minority, but these minority cases, we believe, tell us something about the general theory of population growth. In this way, cycling populations in ecology are very much like planetary orbits for physics.

Suppose that we wish to devise a theory about the abundance of some population of rabbits. Clearly, there are two ways the number of rabbits might increase: births and immigration. Similarly, there are two ways for the population to decrease: deaths and emigration. We can therefore say that the population at any given time is equal to the initial population plus all new immigrants and all new births, minus all emigrants and deaths. Because it is the birth–death process we are most interested in, let's suppose that there are no immigrants and no emigrants. Let's suppose, for instance, that the population of rabbits is on an island and that there is no way on or off this island. The only way that new rabbits can appear is by reproduction, and the only way they can disappear is by death. The number of rabbits at any time, then, is equal to the initial population plus births, minus deaths.

So far, so good. But what can we say about the births and deaths? Well, one thing is clear: in general, the number of births increases with the number of rabbits, and similarly for the number of deaths. For example, there are more human births and deaths in the United States than in Australia (all other things being equal) simply because there are more people in the United States than in

Australia. After all, rabbits come from rabbits, and humans come from humans—living organisms do not spontaneously generate. So let's consider the birth and death *rate*. That is, let's consider the average number of births and deaths per rabbit.

Let's suppose that the population abundance at some initial time t_0 is $N(0)$. At some later time t_1 the population will be equal to the number at the earlier time plus all the births in the interim period minus all the deaths in this same period. Instead of simply speaking of the *number* of births and deaths, it is often more fruitful to consider the *rate* of births and deaths. This is the number of new births and deaths (respectively) in a given unit of time divided by the number of individuals alive at the time. For our present purposes, we are not particularly interested in the fact that the change in population consists of births and deaths. We can combine these to get a single growth rate R. This is the average increase or decrease in the population in a given time interval per individual. We can now express the population size at time t_1 as the growth rate R times the number of individuals alive at t_0. We can calculate the population at later times by iterating this process. The population at some time t can be calculated via the following fundamental formula:

$$N(t) = N(0)R^t. \qquad (1.1)$$

[This simply says that the population size at t time steps is equal to the initial population $N(0)$ times the growth rate R raised to the power of t.]

Equation (1.1) is central to population ecology and describes *exponential* or *Malthusian* growth, after Thomas Malthus (1766–1834). It describes the default situation for populations—how they behave in the absence of any disturbing factors. The important question remains, however, of how external "forces" act on this background state. Do environmental forces affect the per capita growth rate directly, or do they affect the rate of change of this growth rate (i.e., the "acceleration" of the population abundance)? It is our view that it is the latter, and we defend and support this view in the remainder of this book. In arguing for this view, we will take the advice from the preceding section very seriously: special insights may be gleaned from rare cases. Here the rare cases are cycling populations—populations whose abundance varies in a periodic fashion.

1.3 Metaphors and the Language of Science

Now that we have outlined the connection that we're interested in exploring between population growth and planetary motion, let us say a little more about what this connection amounts to and why we expect it to be fruitful. After all, by our own admission, the connection is merely an analogy or metaphor. But such a connection, we suggest, is not trivial.

In poetry, metaphors challenge us to see connections and make associations that are not otherwise apparent. In science, metaphors and analogies play exactly the same role. They allow us to explore connections in nature that were not apparent before. For instance, when James Clerk Maxwell (1831–1879) was trying to discover the fundamental laws of electromagnetism, he relied on an analogy with Newtonian gravitational theory—both gravitation and electromagnetism have inverse square laws, for instance. The result of this analogy was the postulation of a principle of conservation of charge. Without Maxwell's use of such an analogy, the arrival of modern electromagnetic theory would almost surely have been greatly delayed. Indeed, use of analogy is common in science. Another to employ analogical thinking was Charles Darwin (1809–1882). Darwin, it seems, was impressed by certain analogies between economics and biological systems.

The purpose of such metaphorical connections across branches of science is not to find mechanistic similarities. Rather, its purpose is to generate research programs in one area of science that are motivated by similar developments in another area. In

Thomas Malthus (1766–1834)

our case, we use analogies between mechanics and ecology to generate fruitful hypotheses for ecology. Ultimately, these hypotheses need to refer to *ecological* causal mechanisms and must be tested by ecological methods, against ecological data.

1.4   Inertial Population Growth

According to our view, populations are *inertial*. That is, populations tend to grow according to the Malthusian law (i.e., exponentially), and the effect of an external force, such as a change in environment, is to produce an "acceleration"—a rate of change in the per capita rate of change, in the abundance. In effect, this means that population abundance does not respond immediately to changes in the environment. There may be lags in the response. Think of the way an inertial object such as a boat responds to forces: trying to bring a boat to a complete standstill is very tricky because you keep overshooting or undershooting the stationary state. This is because of the delay brought about by inertia—the boat "wants" to do what it was doing before. So, too, with populations.

The primary reason for this inertia in population growth (on the timescale of common interest, several generations), we believe, is the *maternal effect*. This is the phenomenon of "quality" being transferred from mother to daughter, the idea being that a well-nourished and healthy mother produces not only more offspring but healthier offspring. So, an individual from a healthy mother experiencing a deteriorating environment will do better and will be able to continue reproducing better than individuals in the same environment not blessed with a healthy mother. This means that the population abundance at any time is the result of both the current environment and, to some extent, the environment of the preceding generation.

An important test of any new theory is whether it is able to explain what was not explained before, and whether it is able to establish new connections between theories. In chapter 2, we review some of the results that we attempt to explain in following chapters. We also discuss the important question of whether ecology has laws.

Two

Does Ecology Have Laws?

It is often claimed that ecology, unlike physics, does not have laws. In physics we are able to identify various laws such as Newton's law of gravitation and Ohm's law of electricity. These laws state the relationship between two or more quantities: Newton's law specifies the relationship between gravitational force and mass, and Ohm's law specifies the relationship between potential difference, electric current, and electrical resistance. It is possible to state these relationships so precisely because the physical world of massive bodies and electric currents is relatively simple, or so this popular line of thought goes. Ecology, on the other hand, is messy. We cannot find anything deserving of the term *law*, not because ecology is less developed than physics, but simply because the underlying phenomena are more chaotic and hence less amenable to description via generalization.

We disagree with this line of thought. In this book we are concerned with uncovering relationships in ecology, some of which, we believe, deserve to be called laws. Before we embark on such a task, however, we need to see what's wrong with the line of thought presented in the preceding paragraph; otherwise, we would be simply embarking on a wild goose chase. That is, we need to at least make a case for the possibility of laws in ecology. We begin by presenting a number of candidates— the so-called allometries. We then present examples of laws in physics. Then, once we have some examples at hand from both ecology and physics, we turn to the more general question of what natural laws are. The way we tackle this issue may seem somewhat indirect, but it turns out to be rather difficult to give a complete and adequate account of what natural laws are. We therefore look at examples of laws in order to get a "feel" for them, before we discuss what we might expect a general account of natural laws to look like. Then, once we understand more about natural laws in general, we will be much better placed to understand what laws of ecology might look like.

2.1   Ecological Allometries

We believe that there are some very good candidates for laws in ecology. Indeed, in chapter 1 we've already presented one of these—the law of exponential growth. For now, we'd like to give a few examples of some ecological relationships we believe deserve to be called *laws*. These relationships are both important for later developments and of considerable interest in their own right. We have in mind the so-called *allometries* of macroecology (see Calder, 1984, for details; see also a very interesting collection edited by Brown and West, 2000). These are remarkable statistical regularities that hold between various biological and ecological quantities. We present a few of these in the order of their discovery.

Kleiber Allometry

The first allometry was noticed by the biologist Max Kleiber in 1932. Kleiber studied the relationship between the size of animals of various species and their rate of metabolism, and he made a remarkable discovery: basal metabolic rate (i.e., calories "burned" while at rest) is proportional to a 3/4 power of body weight. This means that if one animal is larger than a second by a factor of 10,000, the first animal will have a basal metabolic rate larger than the second by a factor of 1,000. This relationship has now been confirmed with a very high degree of precision and has been found to hold for animals as small as shrews and up to

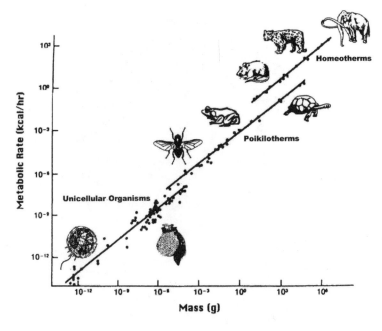

Figure 2.1. Kleiber allometry between body size and metabolism (from Brown and West, 2000, p. 6, modified from Hemmingsen, 1960). Displayed in three functional groups: unicellular organisms, poikilotherms (cold-blooded vertebrates and invertebrates), and homeotherms (warm-blooded birds and mammals). The data for each functional group have been fitted with a line corresponding to $(\text{mass})^{3/4}$. That is, the lines for each group are parallel, having different intercepts and the same slope. Reprinted with permission from Oxford University Press.

the largest animals living. In fact, similar relationships have been found to hold for organisms as tiny as bacteria (see figure 2.1).

It is not surprising that there should be some loose relationship between metabolism and body size; indeed, it is not surprising that we should find that larger animals have a greater base metabolic rate than do smaller animals. After all, larger creatures require more energy input to maintain their various bodily functions. What is surprising is that there should be some relationship

Tom Michael Fenchel (b. 1940)

that holds *across* species and that this relationship should be that metabolism is proportional to 3/4 power of body mass.

That the relationship should hold (though admittedly only approximately) across species is surprising because it would seem possible, at least, that some species would be more or less efficient in using their energy. For example, it would seem possible that an animal the size of a dog could be so efficient in using its energy that its basal metabolic rate would be only marginally higher than some inefficient mouse-sized creature. Kleiber's allometry tells us that there are no cases like this in nature.

John Damuth (b. 1952)

That the relationship should involve the 3/4 power is perhaps the most surprising aspect of Kleiber's allometry. After all, animals take in energy through two-dimensional surfaces (e.g., stomach lining), and this energy is used to maintain three-dimensional bodies. One would thus expect that the relationship of metabolism to body size would be one involving a 2/3 power. Indeed, this is what Kleiber and others expected to find, if anything. Although the data are not so precise as to support 3/4 as the exact power (*no* data are that precise), it is very clear that the power is not 2/3 and that it is close to 3/4.

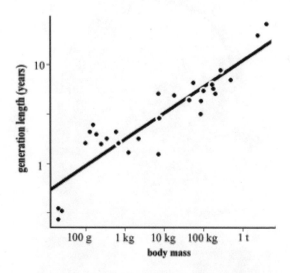

Figure 2.2. Allometry between generation time and body mass (based on data for 29 mammals from Millar and Zammuto, 1983). The equation for this allometry is (generation time) = $1.74 \times$ (body mass)$^{0.27}$.

Generation-Time Allometry

This allometry was suggested by various authors but was made widely known by J. T. Bonner in 1965 in his book *Size and Cycle*. It was noted that the maturation time of organisms of different sizes was also related to their size. More precisely, the suggestion is that maturation time is proportional to a 1/4 power of body weight. This means that an organism 10,000 times bigger than another will take 10 times longer to mature to reproduction age. Others have considered overall longevity and a variety of other life-history timings and obtained similar results (see figure 2.2).

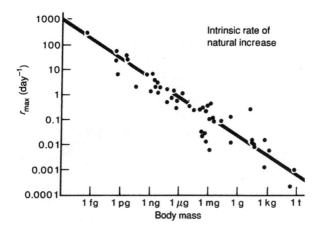

Figure 2.3. Fenchel's allometry between r_{max} (maximal rate of increase) and body mass for 42 species (ranging from a phage to *Bos taurus*, an extinct ancestor of cattle; reprinted from Charnov, 1993, p. 116). The equation for this allometry is $r_{max} = $ constant x (body mass)$^{-0.26}$, where the constant has different values for each group (unicellular, poikilotherms, homeotherms). Reprinted with permission of Oxford University Press.

Fenchel Allometry

In 1974, Tom Fenchel proposed that the maximal rate of reproduction of species was also related to body size. This time the relationship seemed to be that the maximal reproduction rate declines with a power of 1/4 of body weight. So an animal 10,000 times as big as another will only be able to reproduce 1/10 as fast (see figure 2.3).

Damuth Allometry

In 1981 John Damuth noticed that the average density of mammals and birds in their natural environment was also a function of body weight. Damuth found that the density declined as a 3/4

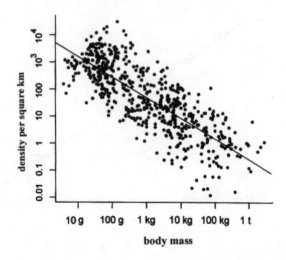

Figure 2.4. Damuth's allometry between body mass (g) and population density (number per km^2) for 564 mammalian species (after Damuth, 1987). The equation for this allometry is density = $10^4 \times$ (body mass)$^{-0.76}$.

power of body size. So, for example, if one mammal is larger than another by a factor of 16, typically there will be 1/8 as many of them per unit area (see figure 2.4).

Calder Allometry

In 1983, William Calder hypothesized that the period of oscillations of animal species' numbers, in those that oscillate, is also related to body size. The relationship is that the period of oscillation is proportional to 1/4 power of body weight. So, for example, if one animal is 10,000 times larger than another, the populations of the first will cycle (if they cycle) with a period 10 times as long as the second (see figure 2.5).

There is also an interesting allometric relationship apparently closely related to the Calder allometry: all physiological and life-history-related timings such as heart beat, maturation time, and the like, are proportional to 1/4 power of body weight. It may

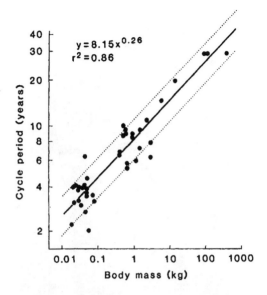

Figure 2.5. Calder's allometry, the relationship between body mass and population cycle period, for 41 species of mammals and birds (from Peterson et al., 1984). Each point represents one species (data for each species averaged for all populations, ranging from 1 to 19). Both body mass and the period scales are logarithmic. Note that a cycle period is expressed here in years, not in generations. Periods of cycles expressed in units of generation times are discussed in section 4.2. The 95% confidence interval for the predicted cycle period is shown. Reprinted with permission of *Science*.

be that the Calder allometry is a result of this more general life-history allometry.

As we've already mentioned, the Kleiber allometry is remarkably accurate over an extraordinary range of body sizes. The others are less accurate, but nevertheless, in each case, there is clear evidence of the relationship described.

It is also worth mentioning a few consequences of these allometries. First, it follows from the generation-time and Fenchel allometries that the lifetime reproduction of typical individuals is

William Calder (1934–2002)

independent of body size. That is, elephants and rabbits produce roughly the same number of surviving offspring over their entire lives. Next, it follows from the generation-time and Kleiber allometries that metabolism per lifetime is proportional to body size. Finally, it follows from the Damuth and Kleiber allometries that the total metabolism per unit of habitat is, crudely, the same across species. That is, the metabolism of a square mile of horses is the same as the metabolism of a square mile of mice.

Perhaps the most surprising thing about all the allometries is that there is a relationship at all. After all, why should there be *any* relationship between metabolism and size or between size and population density? What is the explanation of these allometries? And, in particular, what is the explanation for the recurring 1/4 and 3/4 powers? The bottom line is that no one knows. There are various explanations for the 1/4 and 3/4 powers that range from fractal geometry (Southwood, 1976; West et al., 1999), to physiology, to evolution; but none has managed to satisfy all ecologists. The relationship between the maximum growth rate and body size expressed in the Fenchel allometry is likely to have a physiological explanation, based on a more fundamental metabolism allometry (Kleiber). If reproduction is looked at as any other physiological process, the rate has to be crudely proportional to metabolism per unit weight. This may explain the negative 1/4 power relation of maximum reproduction rate to body size as (body size)$^{3/4}$/body size = (body size)$^{-1/4}$.

It's fair to say, however, that a satisfactory explanation of these and other allometries is something of the "holy grail" of current macroecology. Later in this book, we offer an explanation of some of these, but for now, we continue our exploration of the question of whether ecology has laws. To do this, we need to consider the question of what laws are, in general. So, by way of example, we consider some laws from physics.

2.2 Kepler's Laws

Let's start with a couple of laws from physics that bear similarity to the ecological allometries discussed in section 2.1. The laws of physics we have in mind were discovered by Johannes Kepler (1571–1630). By painstaking analysis of data collected by Tycho Brahé (1546–1601), Kepler recognized certain relationships in the geometry of the orbits of the planets. These relationships are now known as Kepler's three laws of planetary motion:

Johannes Kepler (1571–1630)

Kepler's Laws

1. The orbit of each planet is an ellipse with the sun at one of the foci (see figure 2.6).
2. Each planet orbits the sun such that the radius vector connecting the planet and the sun sweeps out equal areas in equal times (see figure 2.7).
3. The squares of the periods of any two planets are proportional to the cubes of their mean distances from the sun.

As it turns out, these laws hold for any planet, and indeed, they also hold for the moons of planets. At the time that Kepler

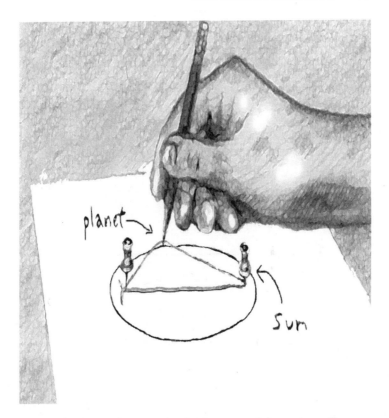

Figure 2.6. Ellipse. The trajectory of a planet around the sun is an ellipse. This is a figure drawn by connecting a fixed length of string to two fixed points (foci). The sun is located at one of the two foci.

proposed these laws, there was no explanation of why they should hold—they were simple brute facts about the way planets and moons behave.

It will serve as a useful comparison to consider another candidate for a law from around the same time as Kepler: the *Titius–Bode law*. This law specifies the distance from the sun of each planet, based on the planet's order in the sequence of planets

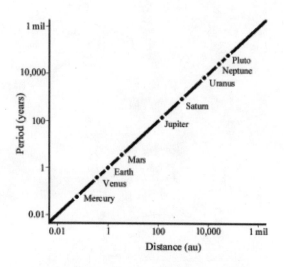

Figure 2.7. Kepler's third law of planetary motion. The squares of the planets' orbital periods are proportional to the cubes of their average distance from the sun. Distances are shown in astronomical units (au), where one au is the average distance of Earth from the sun.

(from closest to the sun to farthest). The main point to note is that the formula for deriving the distances is based purely on observation of what the distances happen to be. The formula has no independent motivation and really is quite ad hoc.

Titius–Bode Law

The distance of each planet from the sun (measured in astronomical units—the mean distance between Earth and the sun) is given by $0.4 + 0.3N$, where N is taken from the sequence $0, 1, 2, 4, 8, \ldots$, and the distance of the nth planet from the sun is calculated using the nth term in the sequence. So, for example, Venus is the second planet from the sun, so N is 1, and the Titius–Bode formula gives us Venus's distance from the sun as $0.4 + 0.3 \times 1 = 0.7$. This agrees pretty well with the actual distance. Indeed, this law holds within a few percent for the seven innermost planets (which were all that were known at the time of Titius and Bode; the law was posited by Titius in 1766). This law also predicted the largest of the asteroids (or planetoid), Ceres, at 2.8 astronomical units.

Note that the relationships described by both Kepler's laws and the Titius–Bode law are only approximate, and neither offers an explanation for the relationship in question. There is, however, a huge difference between the Kepler laws and the Titius–Bode law: Kepler's laws are now underwritten by Newton's theory of gravitation. Indeed, all three of Kepler's laws can be derived from the gravitational inverse square law of Newton's theory. The Titius–Bode law has no such more general theory to underwrite it, and it is now considered no more than an interesting historical curiosity. Indeed, the question of why certain planets are the distance they are from the sun is no longer even thought of as an interesting question—it's seen as simply a matter of initial conditions.

Our main point is that Kepler's laws describe certain relationships that hold for the orbits of the planets and planetary moons. Why these relationships hold is not explained by Kepler's laws—this came later with Newton's theory of gravitation. The Titius–Bode law, at the time of its postulation, shared whatever virtues Kepler's laws enjoyed: it was reasonably (although not perfectly) accurate; it was predictive (it predicted the asteroid belt) although not explanatory. The difference between these two cases shows that distinguishing a law of nature from a mere regularity is no easy affair.

Which of our ecological allometries will achieve the status of a law of nature, as Kepler's laws have, and which will be seen as mere regularities, such as the Titius–Bode law, remains to be seen. We believe that some of the allometries, at least, have an underlying explanation and may be good candidates for lasting, theoretically sound laws of nature.

2.3 What Is a Law of Nature?

There has been a great deal of discussion lately on the question of whether biology and ecology have laws (e.g., Cooper, 1998; Lawton, 1999; Murray, 1992, 1999; Quenette and Gerard, 1993; Turchin, 2001). Unfortunately, a great deal of this discussion has suffered because of a lack of clarity about what a natural law is. As we suggested above, giving a clear account of laws of nature is not easy, which may explain why the important question of what they are has been mostly overlooked by those involved in the debate so far. We are a little more foolhardy than those who have gone before. In this section, we attempt to clarify what a law of nature is.

Let us clear up a few misconceptions about natural laws and the role they play in science. The first misconception about natural laws is that they must be exceptionless. But requiring them to be exceptionless is far too stringent; if we require laws to be exceptionless, there would be no laws, or very few—even in physics. Galileo's law that all massive bodies fall with constant acceleration irrespective of their mass has many exceptions: snowflakes fall quite differently from hailstones and with radically different accelerations. Or consider the law of conservation of momentum: the momentum (i.e., the mass of the system multiplied by its velocity) is constant. In particular, consider the collision of two billiard balls. The momentum of the system, according to the law in question, will be the same after the collision as before. But this is not the case; the momentum of the system after the collision is always slightly less than the momentum before. Or consider Kepler's first law. Not only does this law have exceptions, but *every* planet is an exception. The orbit of any planet is *approximately* an ellipse but because of disturbing factors (e.g., gravitational influences from other planets and changes in mass of the planet and the sun), it is not *exactly* an ellipse. The point is that if natural laws are supposed to be exceptionless, it would seem that there are no laws.

Now it's not too difficult to give an account of why these laws fail. In the first two cases, we've neglected the relevant frictional forces: the effects of air resistance on snowflakes and hailstones; and the frictional forces between the billiard balls and the table. In the case of Kepler's law, we neglected to account for disturbing factors in planetary motions. This suggests that the view that natural laws should be exceptionless can be salvaged if we simply limit the scope of the laws in question. So, instead of the standard statement of the law of conservation of momentum, we limit it to cases where there is no friction. Now the law has no exceptions, but it also fails to be of any use, for the simple reason that there are *no* zero-friction environments. A law, thus construed, tells us nothing about the momenta of billiard balls and the like. In particular, it fails to account for why billiard balls *almost* conserve momenta in their collisions.

The appeal to idealized setups such as frictionless environments and two-body problems seems to be on the right track, however. Such idealizations were called *limit myths* by the philosopher Quine (1960): they are myths because such setups are impossible. Nevertheless, they tell us about what holds in the idealized limit. So a frictionless plane is the limit of a series of real planes, each with less friction than the one before. How such limit myths are to be used in articulating laws of nature is a contentious issue, but it is clear that something like them is needed. Perhaps, as some suggest, laws of nature describe the dispositions physical or biological systems have to behave in certain ways in these idealized setups; in real setups, the physical or biological systems have the same tendencies but the behavior is slightly different because of the interaction of several different tendencies. What is clear, however, is that limit myths are important for our articulation and understanding of laws of nature. In any case, laws of nature (if there are any) are not exceptionless; that's all we're claiming here.

The next misconception about laws is that they should make precise predictions, or, as Popperians are fond of putting it, laws should be *falsifiable*. The idea is that a law L should make some very specific prediction P about what will happen in some setup S. If, in circumstances S, we observe P, then L is (provisionally) confirmed (or at least it lives to be falsified another day); if in circumstances S we do not observe P, then L has been falsified and should be rejected. According to this simple falsificationist line, what distinguishes science from nonscience (or pseudoscience), such as astrology, is that the former but not the latter is falsifiable.

It would take us too far afield to rehearse the many (and in our view, decisive) objections to the simple falsificationist account of science. Suffice it to say that this model fails to account for the holistic nature of confirmation (and disconfirmation), and it finds few supporters among modern philosophers of science. As Quine puts it, "Our statements about the external world face the tribunal of sense experience not individually but only as a corporate body" (Quine, 1980, p. 41). This point was made long ago by Duhem (1954) and more recently by Quine (1980, 1995) and Lakatos (1970). Once we appreciate this basic point about the logic of scientific methodology, it turns out that no hypothesis (or law) is strictly falsifiable in the sense presupposed by simple falsificationism. We can always make adjustments elsewhere in the theory (in what Lakatos called "the auxiliary hypotheses") to accommodate recalcitrant data.

Think of the way in which Newton's law of gravitation was saved from falsification in light of the aberrant behavior of the orbit of Uranus. The auxiliary hypothesis adjusted was the one concerning the number of planets (at the time, thought to be seven). Once an eighth planet (Neptune) with suitable mass and orbit was posited, not only was Newton's law of gravitation saved from falsification, but also the discovery of Neptune was taken by most commentators to be one of the great achievements of Newton's theory. But the simple falsificationist view has a hard time accounting for such episodes because, according to one reading of the simple falsificationists' view, Newton's law was falsified by the orbit of Uranus, and that should have been that—the law should have been rejected. On another reading, Newton's law was not falsified because it could be protected from impending falsification by making suitable adjustments elsewhere. But such adjusting is an option for protecting any law, so it's hard to see how any law could be falsified.

The point we're making here is simply that a single law typically does not make predictions on its own; a great deal of extra theory and facts about initial conditions are required to make any predictions at all, let alone precise predictions. So, for example, although Newtonian gravitational theory makes some rather precise predictions about Halley's comet, for example, it makes much poorer predictions about the trajectories of the smaller asteroids in the asteroid belt (because the latter involves knowing a solution to the intractable N-body problem). Although there's no denying that predictive power in a theory is a virtue, it should not be seen as the sole responsibility of the laws to provide this.

The final misconception about laws is that they are clearly distinguishable from mere regularities. But this is hard to maintain, as the examples of Kepler's laws and the Titius–Bode law demonstrate. One way to try to distinguish laws from regularities, however, is to appeal to explanatory power. The suggestion is that natural laws, but not mere regularities, are explanatory. That is, we assume that appeal to a law will explain the regularity of the events in question. So, for example, Newton's law of gravitation does not merely *predict* the gravitational pull of Earth on the moon, it *explains* it. But this line of thought is also hard to sustain. We all know that explanation must end somewhere, and typically it ends with the laws of nature. In a very important sense, then, such laws do not explain anything—they merely state the fundamental assumptions of the theory.

Reconsider our earlier billiard-ball example. If two billiard balls of the same mass collided such that before the collision one is moving and the other is stationary and after the collision the first is stationary and the second is moving, why is it that the velocity of the second ball after the collision is the same as the velocity of the first before the collision? Because of the conservation of momentum, of course. The law does seem to explain. But this appearance is only superficial. The law of conservation of momentum really just describes the situation; we are none the wiser as to why the two velocities are the same after hearing the story about the conservation of momentum. To see this point from a slightly different angle, consider the question, Why is momentum conserved? We really don't have an explanation of the billiard ball velocities until we have an adequate explanation of the conservation of momentum. It seems, then, that fundamental laws need not be explanatory—indeed, it seems that fundamental laws of nature are an appropriate place for explanation to end and so *cannot* be explanatory.

We take the above discussion to show that whatever laws of nature are, we should not expect them to be exceptionless, we should not expect them to always be predictive, and we should not expect them (in general) to be explanatory or to distinguish cause and effect. This is not to say that they never have any of these features. Indeed, we might even prefer laws that do have some or all of these features. Our point is simply that these cannot be necessary conditions for being a natural law.

We will return to the matter of how one goes about selecting the best theory (or hypothesis) in section 8.2, but very briefly, the matter comes down to choosing theories that exhibit certain aesthetic virtues. The most common such virtue is simplicity, the idea being that if two theories conform equally well with the data, we should prefer the theory that accommodates the data in the simplest way, or is simpler in its underlying assumptions. Although such appeals to simplicity are commonplace in science, they are not easy to justify and they are not without their critics. We will have more to say about such matters in the final chapter.

2.4 Laws in Ecology

Now that we have a better understanding of laws in general, let's return to the question of whether there are laws in ecology.

It seems that a great deal of the dissatisfaction with the candidate laws in ecology is that they are not exceptionless; most laws in ecology are fairly inaccurate in the sense that they have many exceptions, or they only hold approximately. Consider, for example, the Kleiber allometry. The relationship claimed here, although the most accurate of all the known allometries, is only approximate (most data points do not lie exactly on the line in the graph, and some are quite a way from the line). But why should such inaccuracies rule this out as a candidate for a law of ecology? After all, we've already argued that most laws fail to be exceptionless, and it is also very common for laws to hold only in idealized situations. Now, we're not claiming that the Kleiber allometry *is* a law of ecology—just that the fact that it holds only approximately should not rule it out as a candidate.

What else would be required to convince us that the Kleiber allometry, for instance, is a law? Well, for a start, we'd like a story about where and when to expect exceptions. More generally, we'd also like a story about how the exceptions arise. Second, we'd like, although we don't insist, that the law in question be explanatory. That is, we'd like the law not just to tell us that a correlation exists, but to explain why the correlation exists and why it is as it is. To ask for laws to be explanatory may, in general, be asking too much. After all, we've already shown that, even in physics, laws are not always explanatory. There is a sense, however, in which other sciences, including ecology, should be held to a higher standard than physics in this regard. Let us explain.

We've already suggested that, in physics at least, the laws are where explanation might be thought to end. So, for example, Newton tells us that gravitational attraction acts according to an inverse square law (the law of universal gravitation), and this is why planets move in ellipses. But this leaves the law of universal gravitation unexplained, and because the explanation of the planetary orbits rests on this, we don't have an explanation for the latter, either. With the law of universal gravitation, it seems we have reached scientific bedrock—there simply is nothing known to us that is more fundamental than the law of universal gravitation. Now consider a biological example. Let's suppose that we had a law that stated that organisms evolve so as to be well suited to their environment. We should not accept that such a law is fundamental—we should ask why it should hold, if it holds at all. Of course, something like this law is true and is underwritten by Darwinian evolutionary theory. The latter provides the causal story that explains why, when we look around us, we see organisms that are, by and large, reasonably well suited to their environments.

The difference between the two cases is instructive. In physics, unlike biology, ecology, and other branches of science, we are considering the fundamental laws of nature. Insofar as we are considering such laws, we can expect that these laws will not allow, or require, further, deeper explanations. In other branches of science, we are not discussing such fundamental laws (otherwise, we would be doing physics). So, in biology, for example, we expect the fundamental laws will boil down to facts about chemistry and ultimately to physics. This is not to say that we will always be able to *reduce* other sciences to physics. On the contrary, we believe that it is not useful and probably even impossible to reduce biology to physics. What we are claiming here is that *why* questions, if they end at all, end in physics, not in biology. We may choose to stop asking the why questions before we get all the way down to physics (because they have led to another discipline and so they then become the responsibility of that other discipline), or we may be prevented from pursuing these questions further because of the complexity of the subject matter. (Imagine trying to explain the complex interactions in a rainforest in terms of subatomic particles!) For example, why questions in ecology may lead to physiology. Ecologists can therefore rest comfortably with physiological answers to ecological questions. It's not that physiology is in better shape or is better understood than ecology, or that physiology is reducible to physics. It's just that the questions the ecologists began with have led to answers in well-established theory and that's enough.

To sum up this discussion, we believe that there are good candidates for laws in ecology. We have shown that those who would deny that there are laws in ecology may have a somewhat unrealistic account of what laws of nature are and how they operate in other fields of science. Once we set these misconceptions aside, there is no good reason to deny that ecology has laws. At the very least, ecology and physics seem to be in much the same boat in this regard; they both have laws that typically have exceptions, are not necessarily explanatory, may not be predictive, and often invoke idealized situations. Nevertheless, it may be entirely reasonable to strive for ecological laws that have few exceptions, are (in some sense) explanatory, and are not so idealized as to be irrelevant to real-world situations.

Three

Equilibrium and Accelerated Death

If you are an individual belonging to a population that is *at equilibrium*, or at *carrying capacity*, you will not feel too secure. Granted, some individuals may own their own territory or be ahead in the feeding order, and thus will be reasonably well off, but the average member of such a population is in something of a precarious situation. Equilibrium means that births just barely compensate for deaths. Making conditions just a bit worse for an average individual will force the population into decline. It is, however, the case that self-regulatory forces are in action here: if the population size decreases, life per capita gets better, and if the population size increases, life gets worse per capita. This is all assuming, of course, a constant resource stream.

The usual view of this self-regulation is to express it in terms of rises and falls in the birth and death rates, depending on whether the population abundance is above or below the equilibrium. We argue, instead, for a two-dimensional, energy-based view, which is supported by two large-scale observations. That is, we suggest that thinking about equilibrium simply as a balance between births and deaths is too simplistic. Rather, equilibrium is a balance between rates of energy use and rates of consumption, with both birth and death rates a consequence of these metabolic considerations.

3.1 Accelerated Death

For a population to grow exponentially, or with a constant per capita growth rate, an average individual has to produce a constant number of offspring per unit time. This is also true when we are discussing exponential decline. The exponentiality of the process implies the constancy of the physiological state of an average individual. Individuals, of course, use energy at a rate that depends on their specific metabolic rate—this energy is used for maintenance and reproduction.

Let us now consider an extreme situation. Let's assume that there is no food for a given population. Some organisms respond to the absence of food by shutting down their various metabolic functions. They do not die in the absence of energetic input but rather hibernate until food is available again. Simple unicellular organisms such as *Escherichia coli* do just that. Most other organisms, including mammals and birds, die in the absence of food. It is this death process we consider more closely.

If organisms continue to metabolize energy in the absence of food, their internal energy level will decline and their net rate of reproduction will follow this decline. Therefore, we expect the total population abundance to decline with acceleration, not with a constant exponential death rate (curve B in figure 3.1). Figure 3.1 shows the expected behavior of the growth (death) rate and the corresponding curve of the population abundance. Assuming constant decline in the internal energy level, the abundance (in a logarithmic scale) will look like a parabola. That is, the decline will be quadratic in time. The differences between curves A and C in figure 3.1 are a result of the initial reproduction rate, determined by the initial energetic input. In some cases, when the initial energetic input is sufficiently high, reproduction may take place initially, even in the absence of external food.

The main conclusion of this simple analysis is that considering not just the birth and death process, but also its underlying ener-

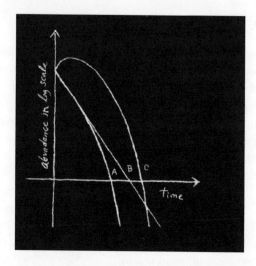

Figure 3.1. Schematic comparison of exponential and accelerated death: the logarithm of abundance declines linearly in the case of exponential death (B) and parabolically for accelerated death. The difference between parabolas (A and C) is due to different initial growth rates.

getic origins, suggests accelerated rather than exponential death to be exhibited by populations in the absence of food.

3.2 Galileo and Falling Bodies

Before Galileo, the accepted wisdom in physics was that heavier bodies fell faster than lighter bodies and that this velocity was constant throughout the body's fall. Galileo refuted this view by considering the following imaginary setup (called a *thought experiment*). (Contrary to popular belief, Galileo did not refute Aristotelian physics by performing real experiments, such as dropping rocks from the leaning tower of Pisa.) Galileo imagined two falling bodies of different weights—the first heavier than the second. He reasoned that if he were to tie these two bodies together with a piece of twine of sufficient strength and allow them to fall freely from the same height, Aristotelian physics would tell us that because the first is heavier than the second, we should expect the first to fall faster until the piece of twine becomes tight, and then the second will retard the first's motion. The velocity of the system (i.e., the two bodies combined) will be *slower* than the first dropped on its own. On the other hand, the mass of the system is greater than the mass of the first body alone and so the system's velocity should be *faster* than the first body dropped on its own. This seems to suggest that Aristotelian physics is inconsistent—it simply cannot be that heavier bodies fall faster than lighter ones. With this thought experiment as his starting point, Galileo then performed some meticulous experiments with balls and inclined planes and deduced that it is acceleration—not velocity—that is constant. Thus, a body that is not falling (because it is suspended from a spring, say) has zero velocity not because the downward *velocity* due to gravity is balanced by the upward velocity due to the spring but, rather, because the downward *acceleration* due to gravity is balanced by the upward acceleration due to the spring. Of course, the velocity is also zero, but it is the balance of the accelerations that is the more fundamental.

The fact that bodies fall with a constant acceleration results in the distinctive shape of the trajectory of projectiles. Imagine standing on top of a cliff and throwing a rock, horizontally over the cliff. If gravity were to provide the rock with a constant velocity, the rock's trajectory would be a straight line. It would have a constant, horizontal velocity component provided by the person throwing the rock, and it would also have a velocity component directly downward provided by gravity. Combining these two components would result in a straight-line trajectory angled downward. The slope of this trajectory would depend on the relative magnitudes of the velocity due to gravity and the velocity due to the throw. But as we now know, thanks to Galileo, gravity provides the rock with a constant acceleration, *not* a constant velocity. This means that the rock has a constant

Figure 3.2. Schematic representation of Galileo's mythical experiment at the Leaning Tower of Pisa. Because of downward acceleration, the trajectories are parabolas, not the straight lines predicted by the Aristotelian worldview.

velocity in the horizontal direction, just as we described above, but in the vertical direction its velocity is *constantly increasing*. The resulting trajectory is not a straight line but a parabola. It starts out horizontal, when the velocity due to the throw is the larger component, and becomes steeper and steeper downward as its velocity due to the acceleration of gravity increases (see figure 3.2).

It is important to note that such trajectories require second-order quantities such as acceleration. Acceleration is the rate of change of the rate of change of the projectile's position; velocity is a first-order quantity, because it is simply the rate of change of the projectile's position. It is not possible to get parabolic trajectories with simple addition of velocities. Similarly, it is impossible to get cycles (e.g., the elliptical orbits of planets) without second-order quantities such as acceleration.

3.3 The Slobodkin Experiment

In the 1980s, Larry Slobodkin conducted some very interesting experiments on water from the Hudson River. He used freshwater polyps, brown and green hydra, to determine the quality of the water from various sites in the Hudson River system. In particular, he placed five of these animals in a synthetic pond of water from the Hudson. In all, there were more than 100 such ponds with water from 51 different Hudson locations. He fed the hydra for 3 weeks before the experiment and then stopped feeding them, recording the number of individuals in each pond on a weekly basis.

The experiment was originally conceived and designed to study how quickly the hydra died in the different Hudson water samples. It turns out, however, that this experiment also tells us something rather important about population dynamics. Because all the hydra populations were without food during the experiment, we have a study of the way in which populations die in the absence of energetic input. According to accepted wisdom in population dynamics, the hydra populations should have died exponentially. That is, if we graph the logarithm of the number of individuals in a given population versus time, we should end up with a straight line. What the experiments showed, however, was that the hydra died with acceleration. That is, the graph of the logarithm of the number of individuals in a given population versus time was, to a first approximation, at least, a parabola—the shape of the flight of a projectile falling under the influence of gravity (see figure 3.3).

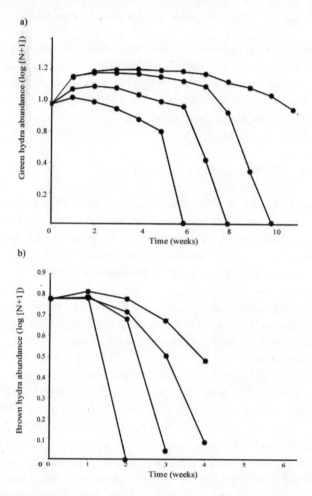

Figure 3.3. Effects of starvation on green hydra (a) and brown hydra (b). These graphs summarize the results of a starvation experiment conducted by Slobodkin (Akçakaya et al., 1988; Ginzburg et al., 1988). The results demonstrate that the decline of populations under starvation is accelerating rather than constant. The population data were combined into clusters, with each cluster based on the longevity of the included populations. The population sizes are shown here in logarithmic scale; the original data in arithmetic scale can be found in the cited references.

Galileo Galilei (1564–1642)

Now, the length of time it took for each of the different populations to die was different, depending on the relevant population's initial growth rate—just as a rock thrown (with the same force) travels farther if thrown at a trajectory of 45 degrees as opposed to, say, a trajectory of 10 degrees. Also, the brown hydra died more quickly than did the green hydra, because the green hydra have extra energy input from their symbiosis with green algae. The important point for present purposes is that all the populations exhibited accelerated death. This, as we've already mentioned, is a telling point against the traditional model of population growth. On the other hand, this is exactly what you would expect from the second-order model we're proposing.

Lawrence B. Slobodkin (b. 1928)

This experiment, although not decisive—no single experiment ever is—presents a serious problem for the standard model of population growth and presents strong evidence in favor of our second-order proposal.

3.4 Falling Bodies and Dying Populations

Although the fact that bodies fall with constant acceleration, not with constant velocity, was discovered by Galileo some 300 years ago, this fact remains counterintuitive. Similarly, accelerated death may seem counterintuitive to biologists who are used to thinking that population decline is caused simply by mortality exceeding reproduction. Although it is undeniable that, when a population is in decline, mortality exceeds reproduction, it misses the point of continuous decline in reproduction (and the increase in mortality) caused by lowered levels of energy available for reproduction and maintenance. In terms of abundance curves, the difference is dramatic—quadratic instead of linear decline (when graphed in logarithmic scales).

The acceleration, not the decline rate, is the fundamental species-specific constant controlled by the rate of the species metabolism. Acceleration (in this case, *deceleration*, because we are talking about declines) is directly proportional to the rate of metabolism. This is because, in the absence of food, an organism's energetic storage will decline with this rate. As storage declines, investments in reproduction and maintenance decline. Thus, the net rate of reproduction will decline, crudely, in proportion to the metabolic rate. In the Slobodkin experiment, the green hydra decelerated more slowly than brown. This is because the symbiotic algae provided some energy—not enough to maintain the population, but enough to slow down the rate of decline of the growth.

Specific abundance curves may differ depending on two initial values: initial abundance and initial internal energy level [or its equivalent growth (death) rate]. Thus, the equation for a dying population is analogous to that of a falling body. It is what is known as a *second-order* differential equation because it is concerned with the rate of change of the rate of change of the population (i.e., acceleration) rather than simply the rate of change of the population. Such equations require two initial conditions before they can give us specific death curves for the population in question.

3.5 The Meaning of Abundance Equilibrium

The standard view in theoretical population ecology is that equilibrium of abundance is a result of the birth rate equaling the death rate. In a sense, this view is correct. In order for a population to have a constant abundance, it has to have equal birth and death rates. This, according to our view, however, is a condition that *accompanies* equilibrium; it is not the *reason* for it. Our suggestion is that abundance equilibrium is the result of the balancing of the underlying energetics of individuals.

Let us return to the above story of accelerated death and imagine that now food comes in at a constant rate of S calories per unit time. With this constant energy input, our population will not die; instead, it will maintain a certain level of abundance dictated by the level of available resources and the population's ability to utilize these resources. If we assume that the incoming resources are simply shared by all individuals in the population, an average individual will consume an amount of S/N calories per unit time (where N is the number of individuals in the population at any given time). But an average individual will metabolize a certain number of calories irrespective of inputs. Indeed, as we showed in section 3.4, the average individual's metabolism is proportional to the acceleration of death under starvation. It follows from this that a population will equilibrate at the abundance inversely proportional to the metabolism rate of a typical individual. Of course, at this equilibrium point, births and deaths will be equal, but this will be a consequence, rather than the cause, of equilibrium driven by energetic considerations.

We can state this more precisely if we employ the same notation as previously for population abundance, $N(t)$, and introduce a new variable, $X(t)$, for average individual quality (energy resources stored inside an individual). The simplest model is as follows:

$$\frac{1}{N} \times \frac{dN}{dt} = f(X),$$
$$\frac{dX}{dt} = -m + k\frac{S}{N}. \tag{3.1}$$

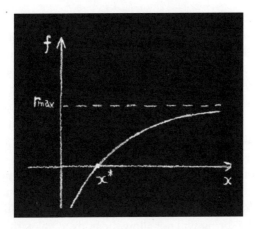

Figure 3.4. The shape of the dependence of the growth rate $f(X)$ (units are 1/time) on the average individual's energetic content X (units are calories). The value of X^* corresponds to a balance between birth and death rates; r_{max} is the maximal possible growth rate for a given species.

Here, $f(X)$ is the net, per capita reproduction rate (birth rate minus death rate), which is a function of the internal resources, or individual quality (figure 3.4). The resources are expended with the metabolism rate, m, and they are imported at a rate proportional to the total available resources per capita. (The additivity of metabolism and consumption is a simplification; the absence of X as an argument changing m and k is another simplification.) At equilibrium, quality X^* is such that $f(X^*) = 0$, or births exactly compensate deaths. Equilibrium abundance is determined by equilibrating the second equation, $N^* = kS/m$. That is, at equilibrium, metabolic losses are exactly compensated by the consumed resources. If the rate of metabolism were constant, independent of X, this model would produce cycles of abundance. Metabolism, however, is known to depend on X (Calder, 1984). This dependence acts like friction and ensures that both quality and abundance equilibrate. Note, also, that the time scale of the processes we are referring to is quite short in comparison with generation time, a more relevant unit of time discussed in chapter 4. For example, mammals without food die in a small fraction of their generation time. Hydra happen to die very slowly in units of their generation time. That is why we were able to see the accelerated death in Slobodkin's experiment. The main source of inertia that we are focusing on in this book is the maternal effect acting on much longer, generation time scales. This is addressed in chapter 4.

It is worth noting that, according to this model, a population reaching equilibrium under steady resource input is mathematically analogous to a body connected to a spring falling under gravity. The spring will not allow the body to fall with a constant acceleration, and the spring stops the fall when the forces balance.

3.6 The Damuth Allometry

One of the most surprising allometries discussed in chapter 2 is a relationship discovered by John Damuth in 1981. Apparently, the per unit habitat abundance of mammals and birds scales with the body size to the power of $-3/4$. Even though the allometry is not terribly accurate (it has an error of up to 100-fold), it is still very clear that there is a relationship here, and moreover, it holds for a range of body sizes of about 100,000-fold. We are now in a position to suggest a simple explanation of this remarkable relationship.

Because equilibrium abundance (with shared resources) has to be inversely proportional to metabolism, Damuth's law is an immediate consequence of Kleiber's law. The view of equilibrium we proposed in section 3.5 has the consequence that a population will equilibrate at the abundance inversely proportional to the metabolism rate of a typical individual. But, by Kleiber's law, the base metabolism rate is related to body weight by a 3/4 power. So, combining these, we get Damuth's law: the equilibrium abundance scales to body size by a power of $-3/4$.

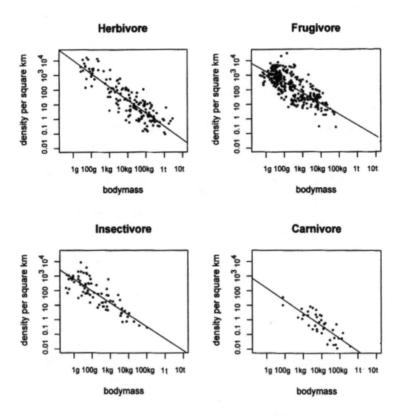

Figure 3.5. Damuth's allometry by species groups. Herbivores are all primary consumers, except for those who primarily eat fruit, which are shown on the frugivore graph. The equations for this allometry by species group are, for herbivore, density = $9 \times 10^5 \times$ (body mass)$^{-0.9}$; frugivore, density = $10^4 \times$ (body mass)$^{-0.74}$; insectivore, density = $4 \times 10^3 \times$ (body mass)$^{-0.82}$; and carnivore, density = $10^3 \times$ (body mass)$^{-0.86}$.

There are a few problems with this explanation. First, we have to assume not only metabolic similarity but also similarity of investment in reproduction and maintenance by a variety of species. But this is not an unreasonable assumption, given the generation-time and Fenchel allometries. Second, we had to assume *perfect* sharing of resources, but in reality territoriality, social structure, and other factors will often result in imbalances in the allocation of resources (think, e.g., of a typical human society). Third, the available stream of resources (S) will vary for different species, and this will affect the result. Carnivores have an order of magnitude less total food available than do herbivores. It is possible that the second issue is an explanation of why the Damuth law is imprecise, and the third may lead to refinements. For example, it turns out that the carnivores' line on the Damuth diagram (figure 3.5) is, in fact, shifted with respect to the ones for insectivores and herbivores, as one would expect.

3.7 A Harder Question

In this chapter, we have considered the case of a starving population. A second-order model attributing population dynamics to the underlying energetic balance gave a reasonable description of two observed phenomena: (1) accelerated death in the absence of food, and (2) the Damuth allometry of typical population density to body size of its individuals.

With the extremes in mind of a starving population (where a population has no food) and Malthusian growth (where a population experiences unrestrained growth with unlimited resources), we will now attempt to address the harder question of the causes and consequences of inertia in population growth. That is, why don't populations respond immediately to changes in conditions? Why are there lags?

Four

The Maternal Effect Hypothesis

In this chapter we describe an effect that produces population cycles and may be thought of as the mechanism for inertia in population growth. This effect, the maternal effect, is the passing of quality (as opposed to quantity) from mothers to daughters. Although the maternal effect has been known at least since Wellington (1957), it has attracted considerable attention since the work of Boonstra and Boag (1987) on cycling voles and a more recent exposition by Rossiter (1991). Still, it has been underappreciated by the theoreticians. This is despite the recent argumentation in its favor by Boonstra et al. (1998).

When mothers are endowed with plentiful resources, they produce not only a larger number of offspring but, apparently, better-endowed offspring as well. It could have been that well-resourced mothers would just produce more eggs, seeds, or children of standard size, or produce a standard number of them with a larger endowment. Neither of these two options seems to be the case. In all plants and animals that have been studied, the pattern is what one may call a *mixed strategy*. Part of the available resource is invested in quantity and part is invested in quality. To the extent that quality investment is made, it produces a maternal effect. With parental care and other behavioral mechanisms, one can easily generalize this to a parental effect, but for the sake of simplicity, we will assume that mothers have full control of the endowment (thus the *maternal* effect). There is also an inverse maternal effect in which adverse conditions negatively affect the quality of the offspring (Rossiter, 1998).

We will use the term *maternal effect* in its most common sense to mean a positive correlation between the maternal environment and the quality of daughters. (See an excellent collection edited by Mousseau and Fox, 1998, for a summary of current thinking on maternal effects.)

4.1 Inertial Growth and the Maternal Effect

The maternal effect will cause inertia in population dynamics. A population growing on a constant flow of resources and growing to its equilibrium value will not stop at that value but will overshoot the equilibrium. The reason is that mothers reproducing when the population is below equilibrium abundance have plentiful resources and their daughters' reproduction responds not only to the daughters' current conditions but also to the conditions their mothers experienced. The same happens when populations decline from a higher abundance to the equilibrium: in this case, mothers were overabundant and thus undernourished. The resulting effect on daughters leads to the undershooting of the equilibrium abundance. Maternal effects can thus easily lead to populations oscillating about the equilibrium value.

A simple maternal effect model assumes nonoverlapping generations and has just two variables: N, the population abundance, and X, the average individual quality (Ginzburg and Taneyhill, 1994; a brief summary of some of the technical details is given in appendix B). Only one parameter is essential for the model: R is the maximum population growth rate of individuals that have very high average quality, and it is assumed to be greater than 1. If this is not the case, the population will rapidly decline to extinction. If t stands for time (in generations), the model has the form

$$N_{t+1} = RN_t f(X_t) \tag{4.1}$$

$$X_{t+1} = X_t g\left(\frac{S}{N_{t+1}}\right), \quad (4.2)$$

where f is a monotonically increasing function of quality, X, and g is a monotonically increasing function of per capita food, S/N_{t+1}. Equilibrial values of the abundance, N^*, and quality, X^*, are easily defined by $Rf(X^*) = 1$ and $g(S/N^*) = 1$.

Note the argument N_{t+1}, not N_t, in the second equation of the model. The reason for this is that, in the absence of the maternal effect, if X_{t+1} did not depend on X_t, the quality X_{t+1} would have been fully dictated by the abundance N_{t+1} at the concurrent generation. Thus, in the absence of maternal effects, the delay is absent and substitution of the equation (4.2) into (4.1) would lead to an "immediate" or direct density dependence. The presence of the maternal effect makes the model fundamentally delayed density dependent, not reducible to the traditional model of direct density-dependent growth. [The latter has the general form $N_{t+1} = N_t f(N_t)$.]

Typical behavior of the maternal effect model is shown in figure 4.1. Population abundance goes up and down with sharp maxima and flat minima. While this is happening, the quality of individuals undergoes a cycle as well. Typical data collection is only for organism abundance and not for the quality of individuals. Thus, what we see in terms of abundance may well be a one-dimensional shadow of the two-dimensional process in which abundance and individual quality interact in joint dynamics.

Note also that the maternal effect is fundamentally connected to the time scale of generations, and this is one of the main advantages of this view of population cycles. Time is always measured in generations in population genetics, because this is the natural unit for describing all evolutionary processes. In theoretical ecology, the time unit is rarely carefully specified. The maternal effect model compels us to use the same time scale as in population genetics—generations.

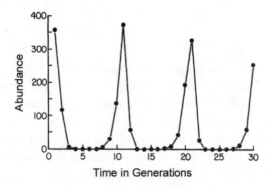

Figure 4.1. Typical behavior of the maternal effect model (equations 4.1 and 4.2) shown in time-series plot of abundance (from Ginzburg and Taneyhill, 1994). Abundance units are arbitrarily scaled; true abundance never reaches zero. Reprinted with permission of the *Journal of Animal Ecology*.

4.2 The Missing Periods

There is one strong prediction that the maternal view of population cyclicity makes. This is that the period length of the cycles is determined by the maximum rate of reproduction of the population (per generation) and that this period cannot fall below six generations in duration. The broken line on figure 4.2 shows the theoretically expected period in comparison with data points. The maximum rate of reproduction, R, has to be above 1.0; otherwise, the population will disappear. When it is slightly above 1.0, the period can be very long; as the maximum rate of reproduction increases, the period of the population cycle decreases. Most observed periods are in the range of 6–12 generations with the maximum reproduction rate, R, of 3–25. It seems reasonable that when a population grows more slowly, it takes a longer time to "go through the motion" of returning to equilibrium than when it grows faster.

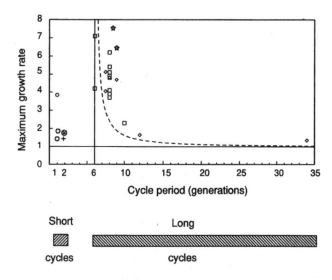

Figure 4.2. The gap between short and long observed cycle periods (in generations) of cyclic species for which the individual quality (maternal effects) hypothesis has been suggested to be the cause of the oscillatory behaivor. The maximum growth rate was also estimated per generations. Squares, forest Lepidoptera species in Ginzburg and Taneyhill (1994); diamonds, northern European voles in Inchausti and Ginzburg (1998); stars, the snowshoe hare time series in Inchausti and Ginzburg (2002); circles, the *Daphnia* spp. from McCauley and Murdoch (1985); cross, moose from Peek et al. (1976); circle with an x, the birth rate of the U.S. human population from Frauenthal (1975) and Easterlin (1961). In the case of voles, two breeding periods per year (spring→fall and fall→spring) are the time units. The broken line corresponds to the theoretical curve of the maternal effect model of Ginzburg and Taneyhill (1994). The straight lines define limits for the growth rate and cycle periods. Only two assumptions are needed to explain these observations: per capita consumption of resources and density dependence delayed by one generation. Both are parts of our maternal effect model (see explanation in section 3.2). The competing view, based on predatory–prey interaction, is more complex and is discussed in section 5.5.

The lower limit of six generations is a bit harder to explain without going into the mathematical details (see Ginzburg and Taneyhill, 1994, for those). But, briefly, recall from elementary calculus that a simple sinusoidal wave has a period of $2\pi \approx 6.28$, and it satisfies a second-order differential equation, $d^2z/dt^2 = -z$. The corresponding discrete second-order difference equation produces an oscillation with a period of 6.0. Or, consider the more general case: $d^2z/dt^2 = -\alpha z$. Here we find that the period is $2\pi/\sqrt{\alpha}$, and so all values of $0 < \alpha \leq 1$ produce periods longer than when $\alpha = 1$. This is approximately what happens in our discrete second-order model, when we write it in terms of the logarithm of abundance. We can therefore say that the limiting period of six generations is a result of the assumption of per capita food sharing, represented in the model as S/N (and resulting in a slope of -1 in the logarithmic scale). The reason is that a hyperbolic decline of $1/N$ corresponds to a slope of -1 in the logarithmic scale. In the more general case, where there is not perfect sharing of food, we would have S/N^α. When there is perfect sharing, $\alpha = 1$; when there is no interference (i.e., consumption rate does not depend on the number of consumers), $\alpha = 0$. The important point to note here is that α has a biologically reasonable range of 0 to 1, and that no matter what value α takes in this range, the period must be longer than six generations. Moreover, if the model is generalized to include damped or unstable oscillations, the period again will increase in relation to the purely oscillatory case. (Indeed, in the mechanical case, this seems intuitively correct, because friction increases the period of an oscillator.) Therefore, the minimal period of six generations is robust with respect to reasonable generalizations of the basic model. Note that the time attribution of the density dependence (N_{t+1} in equation 4.2, not N_t) is both biologically meaningful and necessary for our argument. All of the results depend on this assumption, and it this assumption that also distinguishes our model structurally from other discrete-time predator–prey models.

This minimum period of six generations also accords well with data. Cycles of annual insects are always longer than six years (generations in this case coincide with years). The well-known lynx–hare cycle has a period of about 10 years in duration, or about eight generations of the hare. The 4-year cycle of small mammals (voles and lemmings) is about 12 generations long, considering that these animals have three generations per year, two in the spring to fall period and one over the winter. We consider the simple prediction of the lower limit on periods to be a strong argument in support of the maternal mechanism of population cyclicity.

The maternal effect hypothesis provides a simple mechanism for population cycling that depends only on the population under consideration. It is thus an *internal* mechanism for population cycling, as opposed to external mechanisms such as the traditional predator–prey model. External mechanisms rely on factors external to the single population under consideration. In particular, they rely on the influence of a second population.

There is another well-known internal mechanism of population cycling: the age structure of the population. It is easy to understand this mechanism with reference to what is known as the "baby boom." If there is a bulge in the age structure, overrepresenting the reproductive age category, we expect a generation later to observe a similar bulge, a "baby boom-boom."

In animals, competition for resources by overabundant parents may cause a decline in the number of babies; then that next, smaller generation will have plentiful resources. Thus, we can have a baby boom skipping a generation, or showing a cycle that is two generations long. This is called the *cohort effect*. There is, however, no way that age structure or cohort considerations can account for periods longer than two generations. One can naturally call the baby-boom effect and the cohort effect *first order*, and the delayed effect on the generation time scale *second order*. Second-order effects are necessary for explaining the long cycles, but they are not sufficient. Without a specific mechanism in place, second-order delayed models can produce any periods—including the short ones. Two-generation cycles have been observed in natural populations, but there is no evidence of periods between two and six generations. This gap in the observed periods is strong, indirect evidence for the maternal effect as the mechanism for population cycles six generations and longer (but see the discussion of Murdock et al., 2002, in chapter 5).

Indeed, this may be true of human populations as well. One-generation cycles are seen even in the recent U.S. population history. Two-generation cycles in human abundance have been documented (and are known as the Easterlin hypothesis). There is also some evidence for longer cycles, around 200–300 years (or 12–18 generations; Turchin, 2003b, data from ancient China). But there is no evidence of cycles of any period between. The mechanism for the shorter cycles is, quite plausibly, the cohort effect. The reasons for longer cycles are not known, but the maternal effect, or its financial counterpart (most commonly land inheritance), seems like a good candidate in humans. In prosperous times, when human abundance is low, parents are both physically and financially well endowed. Their children are thus initially better endowed—both physically and financially. For the moment, let's just focus on financial endowment. Let's further (and not implausibly) suppose that each generation of "rich" kids squander their inheritance. That is, suppose that the children spend all their inheritance and do not pass their parents' endowment on to *their* children. On this scenario, we would have a very interesting financial analogue of the maternal effect. This version would also have a single-generational time lag—the only difference is that the parental endowment in this case would be financial and thus more tangible. Obviously, such a resource-related, maternal-effect model would be plausible only in relation to human population cycling. But it may be useful in, and may even shed light on, economic cycles.

Analogies are never perfect. Some, however, are better than others. We feel that the analogy of the uniform motion to exponential growth is very good (chapter 1). The analogy of Kepler laws for planets, as understood at the time of Kepler, to ecological allometries of today is also quite good. In both cases, the mechanisms were and are unknown and a reasonable fit to a simple power scaling law was, and is, quite surprising (chapter 2). In the cases of Galileo's falling bodies and Slobodkin's accelerated death, the analogy is also rather convincing (chapter 3). The alternatives of the uniform rate of falling and exponential death are rejected for similar reasons.

We suggested in chapter 1 another, somewhat tenuous but, we think, useful analogy of cycling populations and planets revolving around the sun. The sun is quite large, and a revolving planet cannot get closer to the center of gravity than the radius of the sun. Therefore, according to Kepler's law, there is a minimal period for planetary orbits (remember, periods decline with distance from the center of gravity). Any planet farther from the sun orbits slower, never faster, than this minimal period, determined by the radius of the sun. Likewise, for a completely different reason (the mechanism of resource sharing), our maternal-effect model predicts a minimal period of cycling equal to six generations of the cycling species. As we explained above, imperfect sharing of resources and other limiting factors can change the period. The change, however, will always result in an increase, so six generations remains the minimum. The mechanism of cycling for planets and populations is, of course, completely unrelated.

4.3 The Calder Allometry

An insightful observation of William Calder (1983) attracted attention to the allometry of the oscillation period of population abundance to body size (see figure 2.5). This allometry, supported by the work of Peterson et al. (1984), and carefully reanalyzed in Krukonis and Schaffer (1991), attributes the relationship of the period to the prey properties in the purported predator–prey pairs. In particular, body size of predators does not exhibit any connection to the observed periods, whereas prey body size does. The original analysis suggested that the period was proportional to 1/4 power of the prey body size. More careful analysis shows that powers vary substantially for various data sets.

Let us try to express the Calder allometry in units of generation times of the cycling species. Generation time is a difficult concept to define precisely for species with overlapping generations. It is clear that smaller organisms have shorter generation times than do larger organisms, but generation time is certainly sensitive, in most cases, to environmental conditions. We have to be satisfied with very crude estimates. Krukonis and Schaffer (1991) reanalyzed the Calder allometry in different ways, suggesting that the slope of the regression line of the original allometry is much less certain than originally suggested. For our purposes, intercepts of various suggested regression lines at the weight of about 1.0 kg will be more informative than will the slopes. The reason is that most cycling species are small, well within the range of 0.1–10 kg, and the intercept at 1.0 kg divided by the intercept of the allometry for the generation time gives a crude estimate of a number of generations per cycle.

Whether or not a particular abundance sequence is cyclic is often a difficult judgment to make, because the usual data sequences are quite short. Krukonis and Schaffer (1991) used four levels of selectivity of the data sets to be included in the cyclic subset. Every level was defined by a more stringent criterion for a sequence to be qualified as cyclic. Independently of the selectivity of data, the period expressed in generations varies on average between six and seven generations (although, not on average, for a specific data series it varies between 3 and 14 generations). If other life history timing measures are used, the result is similar. Calder (1984) estimated the period to be about eight so-called turnover times (standing biomass divided by the annual rate of production).

There is a second part to Calder's discovery, which is of equal or even greater significance to our argument. It turns out that the period of cycling, as a function of predator body size, has zero slope. Body size in this case, just as in the case of prey, stands in for the implied generation time. This finding was reconfirmed by Krukonis and Schaffer (1991). The absence of any positive relationship would not be informative in itself, but it *is* in combination with a strong relationship for the prey. We believe that the two taken together present a strong argument in favor of our view. We are not impressed much by data agreeing with theoretical prediction; we are impressed when theoretical prohibitions are absent in data. This is the second time we have encountered this situation (the gaps in the observed periods being the first instance). Both pieces of evidence, singly and in combination, argue against the predator–prey explanation for the period. This point was originally made by Calder purely on the basis of data, without a specific mechanism in mind.

4.4 The Eigenperiod Hypothesis

According to our maternal-effect–based, inertial-growth view, every population has a tendency to oscillate in abundance with an intrinsic period exceeding six generations in duration. This does not mean that every natural population has to undergo observable periodic behavior. In order to exhibit abundance cycles, the population has to be appropriately disturbed. Let us elaborate with another mechanical analogy.

A suspension bridge or a guitar string is stationary most of the time, but if appropriately disturbed, it oscillates with its own so-called "natural" frequency. In fact, there are many such frequencies, but let us focus only on the dominant one corresponding to the longest observed period. Physicists use the word *natural* for this frequency, stressing that it belongs to the bridge or the guitar string and not to the way they are disturbed. In biology, the word *natural* has so many meanings that we have decided to use the more technical but unambiguous terms *eigenfrequency* and *eigenperiod*. An eigenfrequency is an intrinsic property of the species—its tendency, if appropriately disturbed, to oscillate at a fixed frequency. We stress that this frequency does not depend on the disturbance causing the cyclicity. Now, back to the analogy. Amplitude and the shape of the oscillation that a bridge may undergo do depend on the disturbance; it is only the period that does not. The cause of this simple and clear separation is in the inertial (read: second-order) dynamics that are basic in the world of mechanical objects. Likewise, our view is that inertial effects result in fundamental second-order dynamics and hence the eigenperiod.

The problem of explaining cyclicity of hares in North America or cyclicity of voles and lemmings in northern Europe has been under study for more than 75 years, without a clear resolution (Krebs et al., 2001a). Moreover, the latest experiments (Krebs et al., 1995, 2001b) failed to clarify the issue. In fact, most modern commentators list the competing hypothesis, internal and external, without taking a position on the causes of cycles. It is common to believe that both external and internal mechanisms have to be involved in a full explanation. Our eigenperiod hypothesis does not solve the overall problem, but it seems to explain a major part of it. More specifically, our eigenperiod argument is able to provide a natural explanation for why similar species cycle with similar periods. For example, it is known that hares cycle with a period of around 10 years when lynxes are the major predators, but the period is the same on islands where foxes are the major predators (Peterson and Vucetich, 2002) or when owls are the major predators (Elton and Nicholson, 1942). We hypothesize that the period of 10 years is an eigenperiod for hares. Predators may be the cause of the cycle, and the details of the relevent predator interaction may be required to explain the amplitude of the hare cycle in question—these differ from case to case—but the period of about eight generations (1.25 years is an average generation time for hares) may be the hare eigenperiod.

The so-called 4-year cycles of voles and lemmings (about 12 generations long in this case) may simply be the eigenperiod for these species (Inchausti and Ginzburg, 1998). Note that lemmings cycle in the absence of predators (Turchin and Batzli, 2001) and that this also sits well with our view. What is the chance that the same (hare) or similar (voles and lemmings) species exhibit similar periods in drastically different ecosystems, when the cycling is due to predator–prey interaction? We believe that this is unlikely indeed. [Models based on the traditional view have been constructed (see review in Turchin, 2003a), but, in our view, they have too many parameters to be plausible.]

Our view certainly does not exclude the absence of cyclicity, which is what is seen most of the time for most species. Bridges placed in a viscous liquid will not oscillate. Likewise, strong immediate (not delayed) density dependence will stabilize population abundances. [Boonstra (1994) gives an extensive discussion of the reasons for noncyclicity.] One needs a combination of low immediate density dependence and the right kind of disturbance to cause observable cyclicity. Given that this combination is present, our suggestion is that the period of oscillation is an eigenperiod.

The other commonly observed period of cycling is two generations long. This cycle has a well-understood first-order mechanism of over- and undershooting equilibrium every generation. In some cases this mechanism can lead to cycles that are 4, 8, 16 generations long and even chaotic dynamics (see the classic work by May, 1974a). Note, however, that with noisy data periods of 4, 8, 16, . . . , may appear as a period of 2, because the longer periods remain signwise two generations long: every peak is followed by a trough. Thus, the period of two generations is another eigenperiod explained by the single species nondelayed first-order dynamics. The spectrum of eigenperiods that we suggest consists, therefore, of periods of two generations and more than six generations. This prediction is consistent with the finding of Murdoch et al. (2002).

4.5 What Can Be Done in the Laboratory

It may be even more convincing if inertial growth effects were demonstrated experimentally. Although there are demonstrations of single-species cycles in the laboratory, the periods of such cycles are short in units of generation time of the cycling species. Experiments in the Begon laboratory described in Bjonstad and Grenfell (2001) demonstrate a stable one-generation cycle of abundance. Moreover, with the addition of viruses in one case, and parasitoids in another, the absolute abundances and the shape of the cycles changed but the period did not. This may well be evidence in support of the donor-controlled suggestion (see section 5.4), but the cycles are too short to be the result of the maternal effect.

To clearly demonstrate inertial effects, one has to obtain slow cycles with periods longer than six generations for a single species. This, as far as we know, has not been done. We venture to predict that such a demonstration is possible. In unicellular cultures, larger cells produce larger daughter cells. Thus, the maternal effect has to be present and, according to our view, will cause long-term cycles. With species whose division time is hours to days, the experiment is both practical and achievable. Two properties have to be present to obtain inertial cycles in the laboratory. First, individuals have to die in the absence of food (so, e.g., E. coli is not a good candidate for a test species; see section 3.1). Second, a strong degree of resource sharing has to be present. A low abundance has to noticeably improve the per capita resource allocation compared with the case of high abundance. With these two conditions satisfied, our model predicts that a single species abundance will undergo slow cycling in the absence of any interactions with other species. Moreover, we predict that the *only* cycles we will see will have periods of six of more, or two or fewer, generations.

A numerical study of the long-term abundance data sets can also be helpful in confirming or rejecting our eigenperiod hypothesis (see section 4.4). If we are correct, a "shadow" of this period has to be present, even in noisy data series (Inchausti and Halley, 2001). We therefore predict that a detailed spectral analysis of the long-term abundance data series will reveal a peak at a frequency corresponding to more than six generations of the species in question. To check this prediction is far from easy, considering the relative shortness of the data series that are available and the vagueness of the term *generation time* (NERC, 1999).

We have collected evidence in favor of the eigenperiod hypothesis, although many publications also present counterevidence in favor of the more traditional predator–prey account (the best current summary of this evidence is Berryman, 2003). In most cases, when such counterevidence is carefully considered, our internal forces interpretation of the evidence is as viable as the alternatives. The two suggestions in the preceding paragraphs, a demonstration of inertial effects, and a numerical study of long-term abundance data sets, are our attempt to propose clear tests of the validity of our view.

Although no single experiment or single data analysis would be expected to change people's minds, we believe that demonstrations described above will be extremely helpful in clarifying basic features of population growth—whether or not they support our view.

Five

Predator–Prey Interactions and the Period of Cycling

Since Lotka and Volterra in the 1920s, the predominant view on the cause of cyclicity in population abundance is predator–prey interaction. Three recent books (Berryman, 2002; Turchin, 2003a; Murdoch et al., 2003) develop this traditional line of argument. The predator–prey interaction is certainly capable of producing inertial effects for each of the two species involved in the interaction, and thus can be the cause of the cyclicity. Note, however, that it is the inertia, or the second-order dynamics, that we need for cycles. Second-order dynamics, however, may or may not be the result of interacting species. In chapter 4 we proposed an alternative to the predator–prey model in the form of our single-species inertial model with the maternal effect as the mechanism. We also suggested that the periods of cycling are eigenperiods of a single species (prey in the case of an interacting pair), but that the cause of the cycling may be due to any one (or more) of a number of possibilities. The disturbance that induces the periodic behavior may have its roots in predator–prey dynamics or even higher order ecosystem interactions. All we claim that is internal is the period of cycling, if there is one. This period is an intrinsic property of each population.

To fully appreciate the advantages of our view, it is necessary to discuss some of the details of, and debates about, the predator–prey account of cyclicity. As will become clear, the ratio-dependent idealization of the predator–prey interaction is quite different from the more traditional prey-dependent idealization. Moreover, the ratio-dependent idealization is not conducive to cyclic behavior.

5.1 An Alternative Limit Myth

Ratio-dependent predation is the view that the predator consumption rate is a function of the ratio of food to the number of predators. Although this view doesn't involve as radical a departure from standard theories of population growth as the main topic of this book—inertial population growth—it has, nonetheless, generated a great deal of controversy over the past dozen or so years.

Ratio dependence is easy to restate in terms of an invariance principle: two species' interactions do not change when both abundances are multiplied by the same constant. Such an invariance, *in the absence of other limitations*, seems like a natural and plausible principle. The classical Lotka–Volterra equations are prey-dependent and are, of course, not invariant under such rescaling. The law of mass action, based on a chemical metaphor, lies at the foundation of these equations. Space is thought of as a limiting factor, inseparable from the mechanism of interactions. In the simplest Lotka–Volterra case, multiplying both abundances by 10, say, would produce 10 times more predation. If space or resources for the prey are limited, this is not unreasonable. But space, in our view, is a *limiting factor* acting separately from the pure interactions. It should be treated separately, and it has to be reflected by other terms in the predator–prey model. (In fact, we think that confusing two effects like this is an undesirable consequence of instantism, discussed in section 5.3.) Therefore, we prefer the scale-invariant view of pure interactions. The ratio-dependent invariance may be a "shadow" of a larger invariance that we will call *Malthusian invariance* (discussed in section 6.3).

Let us restate what we have just said in more precise terms. Growth models of two interacting populations of predator and prey are coupled through the consumption rate of the predators. The issue of interest is whether the rate of consumption of an individual predator depends just on the number of prey available (food) or on the amount of food per number of predators. If $f(\cdot)$ is the rate of consumption of an individual predator, the question is,

What does it depend on? In general, because of the interference between predators, it has to depend on both food abundance (N) and the number of consumers (P). Without knowing the exact details of this dependence, there are a couple of idealizations that are plausible. The first idealization is that predator consumption depends *only* on the abundance of food. (The idealization here is that predators do not interfere with one another at all.) The other is that consumption depends on *per capita food availability*. (The idealization here is that the predators perfectly share the resource.) We can represent the situation schematically as follows:

$$f(N) \text{———} f(N, P) \text{———} f(N/P).$$

The true situation is $f(N, P)$, where the nature of the influence of N and P on the predators' rate of consumption is complex; the first idealization (no interference of predators) is on the left; the other (perfect sharing of resources) is on the right. These two idealizations were termed by Arditi and Ginzburg (1989) *prey-dependent* [$f(N)$] and *ratio-dependent* [$f(N/P)$] predation, respectively. The question is not which of these two views is right—for they are surely both idealizations. Rather, the question is which provides the more useful model of real predator–prey populations. In the language of section 2.3, which one is the better limit myth?

5.2 Prey-Dependent versus Ratio-Dependent Models

Most of this debate has been covered in a paper in *Ecology* by Akçakaya et al. (1995), which responded to previously published criticism (Abrams, 1994), and in a review by Abrams and Ginzburg (2000). Here, we address only two points, both related to qualitative differences in the predicted dynamics of the two extremes.

One major piece of evidence in favor of ratio dependence [$f(N/P)$] relates to equilibrial properties of trophic chains in response to fertilization at the bottom. It has been shown (Ginzburg and Akçakaya, 1992; McCarthy et al., 1995), by analyzing data from lakes, that trophic chains respond to an increased reproduction rate at the bottom by monotonic increases at all levels. This increase is close to linear, which is the prediction of the ratio-dependent model. The traditional prey-dependent model, on the other hand, has to invoke many quite intricate modifications to explain this fact. The modifications always increase the number of parameters of the model, and sometimes the number of variables. The ratio-dependent model remains the simplest model to explain the data. (Interested readers may review Akçakaya et al., 1995, for details of the argument.)

The other point, more relevant to the topic of this book, concerns the potential for inherent instability in the interaction between an obligate predator and its prey. In particular, we are interested in the possibility of the obligate predator consuming its prey to extinction, followed by the predator's extinction.

A purely ratio-dependent model has a very narrow parameter range for stable cycling of predator and prey populations. These models are more likely to produce either stable equilibria or what we call a *Gause loop*, the dual extinction described above. This dichotomy is controlled by a simple inequality involving three parameters: prey growth rate in isolation (r), predator death rate in the absence of food (d), and the proportion of prey consumed per unit of time (the so-called attack rate, a). The units of all three are 1/time; and they are all per capita rates. If

$$r + d > a \tag{5.1}$$

predator and prey coexist; otherwise, dual extinction is the outcome. This simple result suggests an experiment: if one can control any one or more of the parameters, one can switch the behavior qualitatively from coexistence to extinction. That dual

Georgyi Frantsevitch Gause (1910–1986)

extinction happens in the laboratory is well known from the time of Gause in the 1930s. Such experiments were repeated by Luckinbill and Veilleux in the 1970s. Luckinbill was even able to produce cyclic coexistence by modifying the thickness of the media in which *Didinium* and *Paramecium* interacted (Luckinbill, 1973). This presumably lowered the attack rate, a, changing the inequality in (5.1) in favor of coexistence. Gause, of course, was unable to avoid dual extinction without refugia for the prey or other forms of reintroduction of the prey. Prey refugia, a portion of the population inaccessible to predators, as a mechanism for turning a Gause loop into stable cycling is the essence of Akçakaya's (1992) lynx–hare model. Akçakaya independently estimated parameters of the model for lynx and hare, and these parameters turned out to be in the neighborhood of $r + d = a$. Within the accuracy range, the model switches between a Gause loop and stable cycling, and with refugia it always cycles.

Instability of well-mixed predator–prey cultures, leading to dual extinction, is a repeated observation in Morin (1999). The traditional prey-dependent models are unable to produce dual extinctions; the prey always avoids extinction because of the structure of the prey-dependent model. The traditional camp, however, appeals to the fact that, at low abundances, differential equations are not really applicable. Thus, even though the prey-dependent model trajectories always escape extinction, dual extinction happens in practice, because populations consist of discrete individuals and because of stochasticity at low abundances. But notice that this is a testable hypothesis. By repeating Gause's experiments with proportionally larger population, one would expect to find differences in the ability to obtain deterministic dual extinction. Such differences would not be demonstrable if the ratio-dependent model is a better descriptor of what is going on. Manipulating initial abundances in relation to equilibrial abundances should help, too. If the prey-dependent model is correct, we should be able to see the switch from dual extinction to coexistence when the experiment is scaled up, and/or when abundances are initially close to equilibrial. This will not happen if the ratio-dependent model is a better description for the pattern.

The results of such an experiment, of course, may not support our ratio-dependent model, in which case, there would be no need for further discussion—the model would be inadequate. Even if the results *do* support our model, there may be plenty of other possible explanations. The same facts can always be accounted for by many theories. Although we concede that both extreme views on predation are imperfect, they do produce two clear and testable questions:

1. Can we systematically switch the predator–prey behavior from dual extinction to coexistence by controlling one or more of the three parameters of the simple inequality above?
2. Can we systematically change the dual extinction to coexistence by enlarging the scale of the experiment (e.g., increase 100 times, from a little vial to a cup size), and by manipulating initial abundances?

We are happy to raise these questions, and we think that experiments designed to answer them will be extremely useful. Of course, such experiments are unlikely to completely decide matters. But they should shed light on which of the two idealizations—prey dependence or ratio dependence—is the more promising.

5.3 The Fallacy of Instantism

The root of the disagreement between prey versus ratio dependence, as is often the case, lies in a seemingly orthogonal, and somewhat philosophical, direction. The issue concerns the interpretation of the rate dN/dt in population-growth equations (Arditi and Ginzburg, 1989).

Most people in the prey-dependence camp (including Lotka and Volterra) take the rate of growth represented by the derivative dN/dt in the relevant differential equations to be truly instantaneous. We call this view *instantism*. When predators reproduce once a year but consume prey every day, the adopted abstraction is to treat the annual reproduction as a corresponding daily rate, and so all rates can be thought of as instantaneous. With this abstraction in place, prey-dependent predation is natural, because in an instant, a single consuming predator may not react to whether there are other predators nearby—the predator in question only responds to the density of the prey.

A common argument for instantism is that every moment reproduction occurs—it does not occur in discrete generational time epochs. For instance, in a laboratory culture, at any given instant there are some cells reproducing, even though the generation time may be about an hour. And in the case of human populations, at any given second there are humans reproducing despite their roughly 20-year generation time. This simple observation suggests that a shorter, or even instantaneous, time scale is the most appropriate. In the case of exponential growth, this argument seems sound because the age structure or stage structure equilibrates and an instantaneous view is a possibility. But as we have stressed repeatedly, the proper subject of ecology is the *deviation* from exponential growth, and from this vantage point, things are rather different. For example, when food supply changes, the response time for bacteria, with a 1-hour generation time, is very different from that of humans, with a 20-year generation time. This common argument for instantism is thus without force in the case at issue.

In our view, the dt in dN/dt should be thought of as a large, finite interval that includes both the reproduction and consumption events. Only at such a scale can we sensibly include, for instance, the conversion of food into offspring. In our view, the scale for defining various rates is different in different cases, but it is never really instantaneous—to assume so is to read too much into the mathematical formalism. We call this mistake the *fallacy of instantism*.

Even in many applications in physics, where differential equations are of fundamental importance, we know that the underlying assumption of continuity is an idealization. Liquids, for example, are *not* continuous; they consist of small molecules. Even space-time may not be continuous, but may come in very small, discrete packets. And in cosmology, Einstein's equation treats the distribution of matter in the universe as continuous, when clearly it is not. For many physical applications, derivatives should thus not be thought of as literally instantaneous rates, but rather as idealizations of finite rates understood at the appropriate scale. In ecology, this point is even more important. Finite time scales certainly force one to consider the importance of predators interfering with one another, and this, in turn, suggests that consumption is ratio dependent.

Skalski and Gilliam (2001) analyzed 19 cases of interactions treated instantaneously. In 18 cases there is noticeable predator dependence—that is, a clear rejection of prey dependence. In 6 of the 18 cases, the more general model was indistinguishable from a purely ratio-dependent one. [A similar result was obtained by Jost and Ellner (2000) when analyzing Veilleux's (1979) experiments.] This is all at the instantaneous time scale. We believe that if the issue is addressed at an appropriate finite time scale, most of the cases would be hard to distinguish from ratio dependence, because the effect of predator interference is more pronounced at the appropriate time scales.

Using long-term data series for 1971–1998, Vucetich et al. (2002) argued in favor of a ratio-dependent model in describing wolf predation on moose. The ratio-dependent model was able to account for 34% of the kill rate when viewed locally and instantaneously. This was substantially higher than other models of equal complexity. At our request, the authors recomputed their results using spatial averages and temporal averages, with a moving window of 2, 3, 4, . . . , years. With the generation time of wolves close to 4 years and moose around 9 years, the results were much stronger in favor of the ratio-dependent view—at about 4-year averaging, the model accounted for about 86% of the kill (figure 5.1). Although the prey-dependent model also performed better with spatial and temporal averaging, the percentage of the kill rate accounted for remained around half that of the ratio-dependent model.

Our view, in fact, calls for discrete difference equations, where time is treated discretely. These equations, however, are notoriously hard to deal with. We therefore continue to use differential equations, but we bear in mind that these are idealizations of the

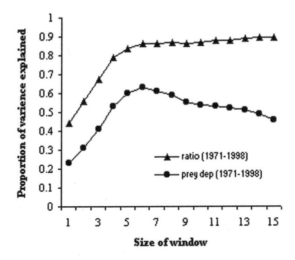

Figure 5.1. Per capita kill rate of moose by wolves as a function of the ratio of moose to wolf abundance (triangles), and as a function of moose abundance (circles), averaged for a number of years shown on the horizontal axis (J. A. Vucetich and R. O. Peterson, personal communication). Instantaneous data are from Vucetich et al. (2002).

underlying finite, discrete-time model. Instantists, on the other hand, treat differential equations literally.

The conflict between these two interpretations of the differential equations is a conflict between two abstractions. Both have their problems. The problems with instantism we've already made clear, but our discrete-time interpretation is not problem free. After all, there are infinitely many different-sized, finite time intervals to choose from, and none, it would seem, is the privileged, "correct" one. When we choose the "appropriate" one, as we've suggested, we too are making an idealization. Which of these two idealizations is preferable? Time will tell; ecology will choose the one that is more useful. We have argued that our view and its natural companion, ratio dependence, is a better idealization than is instantism and its companion view, prey dependence. We admit, however, that the truth is more complex than either of these two idealizations. With current data, it is not easy to distinguish between even the two extremes of ratio dependence and prey dependence. We suggest that adopting the ratio-dependent model is the best default strategy, with the understanding that the long-term goal is to move toward a more general model, when data permit us to do so with confidence.

5.4 Why Period Travels Bottom Up

It is important for our argument that periods do not generally fall below six generations, and that they seem to relate to the number of generations of the prey rather than the predator. If the maternal mechanisms in the prey population are the cause of cyclicity, this is what we would expect to see. We therefore consider the Calder allometry to be strong evidence for this internal mechanism for cycles. The immediate question is, of course: Why doesn't the maternal effect impose its influence on the predator population?

On the one hand, we know that models of predator–prey interactions can produce cyclic dynamics. On the other hand, we have a very simple inertial growth model based on the maternal effect, which attributes the cause of cyclicity to the prey population. This model is also two-dimensional but the dimensions are quantity and quality of the prey, not the abundances of the two interacting species.

There is no doubt that predators and prey interact. If the maternal effect acts in the prey population, causing a cycle, why doesn't it also act in the predator population, causing a cycle of a different period? In general, we have to assume that both species have their own inertial properties, and we would then have to consider a four-dimensional model of predator and prey with the inertial properties of each. The question is why the observed period is close to one driven by only the prey properties.

To explain our argument, we have to consider the so-called *donor-controlled models*. In these models, studied by Stuart Pimm in the 1980s, the prey equation does not explicitly contain the predator abundance as a variable. So, the predator dynamics depend on the donor (prey), and not vice versa. The predator equation does contain the prey abundance as a food source for the predator. The original interpretation of the model is that the predator consumes either old or sick individuals, or very young individuals with a high mortality rate ahead of them. In the last case, the predator is thought of as eliminating the "surplus" production. Such action may have little or even no effect on the dynamics of the consumed population. These have been the narrow interpretations of donor-controlled models.

There is, however, another, more general interpretation of these models, based on the ratio-dependent predation model. In this model, mortality of the prey is expressed as

$$-f(N/P)P, \qquad (5.2)$$

where f is the rate of consumption of the individual predator and depends on the ratio of abundances of prey (N) and abundances of predator (P). The graph of the function f is shown in figure 5.2. For a small amount of per capita food (N/P), consumption rises with supply; it then slows down and levels off when supply is very high. We may consider the increasing, linear part of the curve, f, as proportional to the ratio of N/P. Because there are P consumers, the net result in equation (5.2) is that for low N/P, total predator-imposed mortality is proportional to N. Thus, the model will appear to be donor controlled, with an additional constant mortality imposed by predation. The additional mortality, however, reduces to zero in the absence of predators. The effect of this relationship is that equilibrium abundance of the prey population will be depressed by the presence of predators but the dynamic properties will not be affected. If the

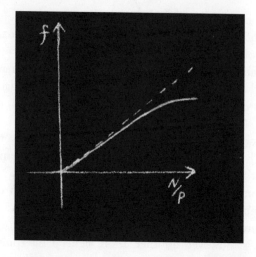

Figure 5.2. Generalized donor-controlled model (the dashed line) as a linear approximation of the ratio-dependent model at low levels of per capita food.

prey population cycles, based on its internal causes, the predator population will follow, simply driven by the prey cycling, with about a generation-time (of the predator) delay.

An assumption we need for this argument to work is that predators are far from satiated even in the best circumstances; that is, the ratio of abundances is such that more food per capita elicits a nearly proportional average individual predator response. The hypothesis is shown in figure 5.2 by the dashed line. We have also analyzed a full four-dimensional system of predator–prey interaction under more general conditions in its linear approximation around the hypothesized equilibrium. The frequencies of the full four-dimensional linear system vary continuously with the deviation of f from linearity. That is, if equilibrium abundances are not too far from the linear range of f, the cycling will appear as though driven entirely by prey dynamics.

So, the answer to the question is not straightforward, but nevertheless interesting. Assuming ratio-dependent predation and the lack of predator satiation, prey abundance is, on average, depressed by the presence of predators, but the cycle period is driven by the prey. This may explain the Calder allometry for cyclic species. [This may also explain a famous experiment (Krebs et al., 1995, 2001b) in which excluding predators failed to eliminate the hare cycle (or change its period) but demonstrated an effect on the average abundance and amplitude of the cycle.]

Another question might be raised at this stage. According to our view, when there is a cycling predator–prey population, the cycle is driven by the prey. Why not, then, attribute the cycle of the prey to cycling of their resources (whatever they may be), and so on down to the bottom of the food chain? It might seem as though we are commited to the view that all cycles are driven by cycles at the bottom of the food chain. This, however, is not right. Not all populations cycle. In fact, as we've pointed out before, cycling is rather rare. Our position is that whenever a population cycles, its period is determined by its eigenperiod. But this, of course, does not mean that any given population must cycle. Recall the analogy with a suspension bridge or a guitar string: when a bridge oscillates, it does so with a fixed period, determined by its eigenvalue—irrespective of the mechanism for its initial movement—but nothing about the eigenvalue of suspension bridges implies that any particular suspension bridge must swing—bridges can, and do, remain stationary. So, in short, we *are* committed to the view that when we find a cycling predator–prey pair, it's the cycling of the prey that drives both—the predators are just along for the ride. But this does not imply that the cause of the cycling goes all the way down. The cause of the cycling goes down only as far as the lowest species in the chain that is experiencing population cycles. And this, because of the rarity of cycling, is typically not very far.

You may have noticed that in explaining both the Damuth allometry in chapter 3 and the Calder allometry in chapter 4, we had to base our arguments on ratio-dependent consumption. It was the assumption of perfect sharing of the resource that allowed us to deduce observed relationships. Ratio-dependent predation is certainly an imperfect model, but it is, in our judgment, superior to the other extreme, the prey-dependent model.

5.5 Competing Views on Causes and Cyclicity

The main competing explanation for population cycles is the predator–prey model. This model suggests that cycles are the result of an interaction between a predator and its prey such that an increase in one is either the cause or the effect of a decrease in the other. This well-entrenched, traditional view goes back to the original work of Lotka and Volterra in the 1920s. This model has a great body of theory and a large number of current supporters. Although it may be the correct explanation of population cycling in some cases, as we discussed in chapter 4 and elsewhere, we do not think that it is the *complete* explanation of *all* population cycles. We thus stress the deficiencies of this traditional view.

The first problem with the predator–prey model is the large number of parameters that the most sophisticated models require and the inability of such models to exclude unobserved behavior. Predator–prey models, in their simplest form, have four parameters (values that have to be determined from data to run the model). The more sophisticated current models have 11–16 parameters. With all this "parametric power," the theory cannot reject any arbitrary periods, including the ones between two and six generations (a.k.a. the missing cycles). In short, the theory is sufficiently flexible to produce any period. But this means that it can't be empirically tested by period data, because it is consistent with *all* possible periods. The maternal-effect hypothesis, on the other hand, makes very specific predictions: no cycles with periods between two and six generations. Moreover, the fact that no such periods have been observed counts in favor of the maternal-effect hypothesis. (The temptation to employ large numbers of parameters in one's model in order to achieve flexibility is sometimes referred to by modelers as *overfitting*. In a joke attributed to Einstein, he allegedly said, "With five parameters I can draw an elephant, with six, it will wag its trunk.")

The second problem for the predator–prey model is the arbitrariness of the choice of prey and predator. For example, lemmings undergo oscillations of a period similar to that of voles (4 years), and given the similarity of these two herbivores, one would expect a similar explanation of the cycles. The standard predator–prey explanation for voles is that voles are the prey in a predator–prey pair. This won't work for lemmings, though, because lemmings on islands are known to be without predators and yet still undergo a 4-year population cycle. This does not perturb the predator–prey theorist; they simply construct a model in which lemmings are the *predator* and vegetation is the prey. Apart from the obvious charge of being ad hoc, there is an important sense in which this model is not explanatory. Because the mechanism in each case is quite different, the account fails to explain the similarity of the cycles for lemmings and voles—it's just viewed as a coincidence that the two cycle with the same period. One would hope for a more satisfying account—one that respects the intuition that it's no coincidence that both cycles are typically 4 years long.

To obtain cycles, one needs a model that is at least second order; it does not necessarily have to be a two-species model. A second-order model can result from either internal mechanisms (e.g., the maternal effect) or external mechanisms (e.g., predator–prey interactions). It is thus the more subtle properties of population cycles—such as the presence or absence of gaps in the periods—and not the mere existence of cycles, that must decide whether second-order models based on internal or external mechanisms are to be preferred.

Murdoch et al. (2002) analyzed periodic data series, separating them into two classes: specialist and generalist predators. The analysis produced a clear gap in periods, with most generalists exhibiting cycles with periods of two to four maturation times. Most life-history data are available in terms of time to maturity, or overall longevity. These differ by a factor of 2.5 for mammals, for example (Charnov, 1993), with a harder-to-define generation time somewhere between, closer to the time of maturity than to the total longevity. Because maturation time is less than generation time, we may attribute this class to two-generation cycle category. The most interesting finding of Murdoch et al. (2002) is that specialist predators cycle with very long periods, certainly more than six maturation times of the prey, but possibly much longer. There is a clear gap in periods between generalists (short) and specialists (long). This gap is clearly observed when the period is appropriately scaled.

Murdoch et al.'s interpretation of this finding is that specialists' cycles are driven by predator–prey interactions whereas generalists "average" various sources of food and are driven by their own density dependence. We agree with the latter but question the former of the two interpretations. Our view of the maternal effect causing oscillations of the prey and predator following only the dynamics of the prey is equally plausible. In fact, considering the Calder allometry for the periods, it seems even more likely. Also, it is not clear that species can be uncontroversially separated into the two classes in question (specialist and generalist). All that can be said with certainty is that there is a bimodality of the period distribution.

The true cause of the cycles of specialists will not be determined by comparing equations. Some predator–prey models are capable of producing the gap in periods just like those of our simpler, single-species view. Neither can be proven correct based on equations alone. It is the total evidence and the simplicity of the argument that moves us toward the view outlined in this book. But it will be manipulative experiments that will someday resolve the issue.

If we turn out to be correct, the title of the Murdoch et al. (2002) paper, "Single-species models for many-species food webs," will be applicable to both generalists and specialists. The authors, of course, meant to attribute this only to generalists. The old controversy over the causes of cyclicity in nature seems to have heated up again in the last few years, but this time on a new level, with more data and a better set of theories to choose from (Beckerman et al., 2002, is a good reference for intraspecies mechanisms).

Another attempt to explain the period of specialist cycles through a simple, traditional predator–prey theory is made by Schaffer et al. (2001), with the argument reiterated in Turchin (2003b). The period of the classical Lotka–Volterra cycle in its linear approximation is $2\pi/\sqrt{r\mu}$, where r is the growth rate of the prey alone, and μ is the death rate of predators in the absence of food. Because $2\pi \approx 6.28$, if we really had $\mu \approx r$, this period would have been $2\pi/r$, where $1/r$ is the so-called turnover time of the prey, correlated to a generation time. This is the essence of the argument.

We believe that the argument is quite weak on a number of fronts. First, we believe that the Lotka–Volterra model is not an appropriate simplification of the predator–prey interaction, and we have covered the subject in detail in sections 5.1–5.3. Even if one still accepts the Lotka–Volterra model as a basis (as many would do), there is still no reason why μ has to be related to r. Even if we ignore our argument in chapter 3 about death in the absence of food being accelerated, and we assume exponential death (as in Lotka–Volterra theory), there is no known reason to even approximately equate μ to r. This seems to be a move based more on a desire to get the right answer than on any evidence.

It would be premature to pass final judgment on the causes of all cyclic dynamics in population ecology. Indeed, the predator–prey model is not without merit. It does seem implausible, however, that predator–prey models are the best account of all population cycles of six generations or longer. We believe that in many, if not most, cases, the maternal-effect model is the better account.

We conclude this chapter with a reminder about the nature of the phenomena here. Populations cycle by missing the target—the equilibrial level dictated by resources. They would grow exponentially if the resource limit was not present, but they are pulled toward the equilibrium at both high and low abundances. This is not unlike planets, which were it not for the sun's gravitational field would fly uniformly off into space. Instead, the sun's gravitational field pulls them back, and, if conditions are right, they settle into stable orbits around the sun.

Six

Inertial Growth

There are two major difficulties in incorporating maternal-effect ideas into practical descriptions of population dynamics. Both relate to the usual way in which population data are recorded and made available to theorists.

The first difficulty is that the data are available only as numbers of individuals, or as some index of abundance, for a given census area. There are no concurrent data on individual quality or its correlate, say, average body size. Therefore, if we wish to check the inertial hypothesis based directly on the concurrent changes in abundance and quality, we are out of luck. We only have one of the required data sequences.

The second difficulty is that, except for a few cases of biannual counts, typically data are available only as annual counts. Because the maternal effect is attached to the mother-to-daughter transfer, it is linked to the generation time unit (which is generally not one year), whereas data come in a chronological sequence. The exception includes annually reproducing species for which one year is the generation time. This is one of the reasons why cases such as the Lepidoptera described by Ginzburg and Taneyhill (1994) are so useful in defending the maternal-effect model.

6.1 The Implicit Inertial-Growth Model

The term *inertia* was first used in relation to population growth by William Murdoch (1970). He interpreted the term generally as a property of overshooting equilibria due to any one of a variety of causes, without a reference to a specific, single cause. In this chapter, we adopt Murdoch's sense of inertia and try to flesh out the details of a general account of inertia in population growth that is able to accommodate a number of possible mechanisms for the inertia. To motivate this, however, we return to the two difficulties mentioned above.

Our approach to these difficulties—annual data and absence of quality parameters—is to address both simultaneously. We will employ a continuous-time differential equation with dt in dN/dt standing for 1 year. Differential equations have been used from the very start of theoretical developments in ecology by Lotka and Volterra. The unit of time did not matter for most of the history of the field, because no one ever seriously tried to estimate numerical values of the parameters. The models were, and still are, commonly used to produce *qualitative* predictions. This trend has been noticeably reversed since the 1990s, when models using chronological time began to be developed. These latter models employed estimated parameters that reproduce major features of cyclic populations. Although most authors used the predator–prey mechanisms as the basis for their models, a parallel development produced similarly fitting descriptions using discrete, generation time steps (rather than chronological time steps) based on the maternal effect.

It is a well-known mathematical fact that a system of two first-order differential equations, written in terms of two variables, is equivalent to a single second-order differential equation, written in terms of one of the original variables. (There are minor restrictions on this theorem that need not concern us here.) A well-known Hamiltonian form of Newton's laws, for instance, presents them as a system of two first-order differential equations, rather than a single acceleration-based second-order model. It is traditional in ecology to start with two first-order dynamic equations. In the opinion of some, this is the only way to clearly describe a mechanism. In this chapter, we attempt to write a single second-order model in terms of abundance only (in logarithmic scale, as always) but representing an unknown (or hidden) second variable. It will be important to see that this attempt does not relinquish the generality of the more familiar system of two rate-based equations.

The approach we are suggesting here does not depend on the precise nature of the second, unobserved dynamic variable in joint dynamics with population abundance. It can be an unobserved predator or prey abundance, or the unobserved and changing intrapopulation average individual quality. Inertia can be caused by both mechanisms. The existence of this second variable simply tells us that we have to describe a two-dimensional process, with abundance being one of the dimensions. Our choice in this situation is simply to write a general second-order differential equation and view the abundance and its rate of change as two independent variables. We will call such a second-order model, written in terms of abundance and its rate of change, an *implicit inertial growth model*—it is implicit in the sense that the second variable is not explicitly representing any particular mechanism for the resulting dynamics. However, if predation is known to be the mechanism, the growth rate of the observed population can stand for the underlying unknown predator abundance. If the mechanism is known to be maternal effect, the growth rate can stand for the unobserved individual quality.

The description of a fundamentally discrete process of population growth by a continuous differential equation is a serious issue that we have already broached but will briefly revisit here. The solutions of differential equations exactly coincide with the solutions of the corresponding discrete models only in the case of linear models. Not all of the models we use, however, are linear. There is thus a substantial issue here: in what sense do the easier-to-study differential equations correspond to their discrete analogies? This is an open and unresolved question. In very general terms, the simplicity of the continuous model may be misleading. Chaotic dynamics appear in discrete models in a single dimension (May, 1974a), but we need at least three dimensions to obtain similar effects in the continuous case.

One argument, commonly given for the validity of the continuous description, appeals to overlapping generations. In most populations, because of a distribution of developmental stages, there are always some individuals reproducing. Such almost-continuous features of reproduction are clearly important and yet would be left out of any discrete model. The only viable alternative, it would seem, is to invoke the idealization of a continuous-time model. But as we've already pointed out (in section 5.3), this argument is clearly applicable in the case of Malthusian growth when age or stage distribution is stable, but when our focus is deviations from exponentiality, the argument does not seem so convincing. Moreover, the argument is completely unconvincing when reproduction is synchronized, as in annual species. Although we acknowledge that this traditional argument for continuous description is not without some force, we have to admit that our use of differential equations in this chapter has to be viewed as an approximation with unknown accuracy. It is our, and most ecological theoreticians', hope that we are crudely describing what goes on in the underlying discrete-time process, at least for some finite duration of time.

The equation we need in the general situation described above is

$$\frac{dr}{dt} = F(n, r), \quad (6.1)$$

where $n = \ln N$ (abundance measured in the logarithmic scale), t is time in years, $r = dn/dt$ is a per capita growth rate, and F is the acceleration of the logarithmic abundance or the rate of change of the growth rate, dr/dt.

The model obviously produces exponential growth [or linear growth of $n(t)$] when $F = 0$. F therefore characterizes the deviation from Malthusian growth. This deviation depends on both population abundance, n, and the growth rate, dn/dt, which, recall, stands for the unmeasured, hidden variable engaged in joint dynamics with abundance, $n(t)$.

Predator–prey models have always been written in terms of differential equations, both because of technical convenience and because it is not clear how to write a reasonable discrete-time model in any other way (because generation times of predator and prey can be very different). The unit of time for these models is, in principle, arbitrary, but it is commonly thought of as annual, unless seasonality within the year is incorporated. If seasonality is incorporated, the unit of time would be days or weeks. This more specific time scale, in the opinion of some authors, adds power to the theory. We remain unconvinced.

Maternal-effect models are written in units of generation time. If viewed in chronological time, the change of units leads to a simple transformation. If T is generation time of the species (in years) but we wish to write an equation on an annual time scale, we need to reexpress rates per generation in terms of rates per year. The rate per generation is simply T multiplied by the rate per year. For acceleration, dr/dt, we have to multiply the annual acceleration by T^2 to move to the generation timescale. Obviously, the reverse is also true: to reexpress rates per generation in terms of rates per year, they have to be adjusted accordingly. For example, a rate of growth of 2.4 times per generation will be 0.8 per year if the generation time is 3 years; acceleration per year, in this case, will be nine times smaller than the one measured per generation.

Let us now describe the span of population dynamics events that a good general theory needs to cover. One extreme of the spectrum is extinction in the absence of food, which is a negatively accelerated process, as described in chapter 3. In this case, $F = \text{const} < 0$. On the other extreme is exponential growth with the maximal possible growth rate, r_{max} (where r_{max} is species specific). This is an ideal exponential growth, with a rate that cannot be exceeded.

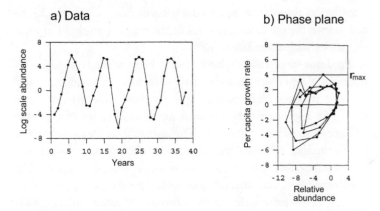

Figure 6.1. Population dynamics for *Zeiraphera diniana* (larch budmoth; Ginzburg and Inchausti, 1997): (a) time series data; (b) phase plane representation of the time series. Reprinted with permission of *Oikos*.

The dynamics of the implicit inertial equation is naturally represented in the abundance–growth-rate phase plane. This is simply the space of various possible values for the abundance and growth rate. In this representation, growth rate is an independent variable standing for a hidden, unobserved dimension (which might be quality or it might be abundance of another species). Figure 6.1 shows such a representation. Note that the cycle of the population looks asymmetric: it grows much more slowly than it declines. That is, it takes longer to grow to the maximum than it does to decline from the maximum to the minimum (Ginzburg and Inchausti, 1997). The reason for this is the existence of the maximum rate of increase, r_{max}, but no such bound on the rate of decrease. Organisms can die instantaneously, but they cannot reproduce faster than a certain species-specific rate. Indeed, the very existence of r_{max} turns any unstable equilibrium into a limit cycle. There is no need for other explanations of why the species in question does not, as it were, spiral out to extinction.

Direct density dependence or limiting factors act analogously to friction in physics. The presence of these will dampen the oscillations while elongating their period. Oscillations in the case of a perfectly constant environment will die out, and the population abundance will equilibrate. However, in a noisy environment, populations will exhibit correspondingly noisy trajectories but with their natural frequency of oscillation evident through the noise. This is not unlike a pendulum subject to random hits. Its trajectory is random but the natural eigenfrequency dictated by the pendulum characteristics will be clear in its behavior. We therefore expect, based on the maternal-effect model, that in general, noisy population trajectories will reveal oscillatory tendencies with the dominant period exceeding six generations. Indeed, there are suggestions of such periodic behavior in many data series, but the series in question are typically too short for any definitive cyclic behavior to be established.

We will treat r_{max} as an ideal maximal growth rate that can be approached infinitely closely from below, but cannot be reached. This treatment allows us to constrain the expression of F by a simple condition:

$$F(n, r_{max}) = 0 \qquad (6.2)$$

irrespective of the value of n (where $n = \ln N$). This condition disallows the growth rate to increase past the value r_{max}, so that all the events in population growth are happening in the portion of the n–r phase plane with arbitrary values of n but restricted values of $r < r_{max}$ (figure 6.1). Between the two extremes of accelerated death and the maximal exponential growth lie a variety of dynamic behaviors, including equilibration of abundance, oscillatory dynamics, limit cycles, and so forth. Before demonstrating how our model (equation 6.1) describes these behaviors, we address a specific parametric expression of the model that we will use to fit the data.

6.2 Parametric Specification

It is commonplace throughout science to prefer simpler theories and equations. In particular, linear equations are prefered whenever they are up to the task. Linear functions are easier to work with and, in an important sense, are no more complicated than is required to fit certain well-behaved data sets. If linear functions are not up to the task, we look for the next best thing—quadratic functions. More generally, when dealing with differential equations, we prefer, whenever possible, to stick to the lower order terms in Taylor-series expansions (the linear and quadratic terms). So here, in our attempt to "deduce" the equation we need, we will use linear functions, if possible, and quadratic ones when necessary. This is in keeping with many other ecological applications of differential equations (e.g., the logistic and Lotka–Volterra equations). Indeed, Taylor series approximation has often been suggested as a way to "deduce" both the logistic and Lotka–Volterra equations. The difference in our case is that we have a function, F, of two arguments, n and r, and we therefore have to approximate in two dimensions simultaneously.

We will assume an equilibrium point at $n = n^*$ and $r = 0$. We will express the equation in terms of the deviation of the population abundance and growth rate from its equilibrium values. Our suggested equation is

$$\frac{dr}{dt} \approx \left(1 - \frac{r}{r_{\max}}\right)\left[-\alpha(n - n^*) + \beta r\right]. \qquad (6.3)$$

Here α and β are parameters. Equation (6.3) is a quadratic approximation in r, and linear in n, to an unknown function that becomes linear when growth rates are sufficiently small (i.e., when r is much smaller than $r_{\max}$). In this case, equation (6.3) is simply a linear second-order differential equation written in the logarithmic scale of population abundance. The additional term, $(1 - r/r_{max})$, makes the equation quadratic and guarantees that our condition that the growth rate does not exceed r_{max} is satisfied. It would be impossible to reflect this biologically important constraint in the linear equation form; the quadratic form has the minimum required complexity to capture this important feature.

Our equation has three parameters, r_{max}, α, and β, and it requires two initial conditions for the abundance and the growth rate at time zero. The second initial condition, r, represents the value of the hidden variable interacting with abundance in joint second-order dynamics.

Here we can think of α and β as representing the strength of delayed density dependence (α—not unlike the strength of the restoring force in Hooke's law in physics) and the strength of direct density dependence (β—not unlike friction, when negative and "antifriction" when positive, in the mechanical analogy). In the current context, β can be thought of as representing *growth dependence*, the effect of the growth rate on acceleration. This, of course, is equivalent to the usual nondelayed density dependence when $\alpha = 0$. When $\alpha = 0$, equation (6.3) integrates to the usual first-order model.

Figure 6.2 demonstrates a variety of dynamic behaviors covered by our minimalistic, implicit, second-order model. In all cases, parameter values and initial conditions were fitted to produce behavior closely resembling the data. When one is allowed to fit parameters to data, it is not impressive that the fit is achieved. The impressive part, in our judgment, is that the inertial growth equation is very simple and remains the same across very different cases. In short, our implicit, second-order model can capture most of the dynamic complexity seen in real data, but it does so in a simple and well-motivated fashion.

We believe that the inertial growth equation is a plausible candidate to replace the outdated and much criticized logistic equation as the simplest model for single-species dynamics. This model, like the logistic equation, is not really based on any

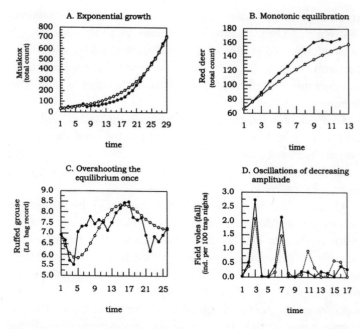

Figure 6.2. Growth curves (solid circles, solid lines) observed in natural populations and described (open circles, dashed lines) by the implicit inertial equation. The predicted population growth curves were obtained by numerically integrating equation (6.3). Parameters and initial conditions used are as follows: (A) Exponential growth (muskox, *Ovibos moschatus*; data from Spencer and Lensink, 1970): $N(0) = 32$, $r(0) = 0.11$, $\alpha = 0$, $\beta = 0$.
(B) Monotonic approach to equilibrium (red deer, *Cervus elaphus*; Clutton-Brock et al., 1987): $N^* = 160$, $\alpha = 0$, $\beta = -0.15$, $r(0) = 0.15$, $N(0) = 60$, $r_{max} = 5$. (C) Overshooting the equilibrium once (ruffed grouse, *Lagopus scoticus*; Middleton, 1934): $N^* = 1921$, $N_{over,max} = 4,860$, $\alpha = 0.09$, $\beta = -0.14$, $r(0) = -0.61$, $N(0) = 1,021$, $r_{max} = 5$.
(D) Dampening oscillations approaching equilibrium (field vole, *Microtus agrestis*, fall census; Hörnfeldt, 1994): $N^* = 0.13$, $\alpha = 2.47$, $\beta = -0.05$, $r(0) = 2.5$, $N(0) = 0.025$, $r_{max} = 3.5$.

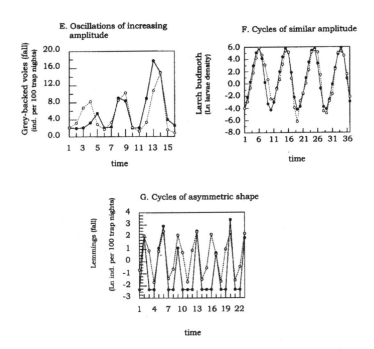

Figure 6.2. (*Continued*) (E) Oscillations of increasing amplitude (gray-backed vole, *Clethrionomys rufocanus*, fall census; Henttonen et al., 1987): $N^* = 4.07$, $\alpha = 1.57$, $\beta = 0.13$, $r(0) = 0.01$, $N(0) = 2.07$, $r_{max} = 10$. (F) Cycles of similar amplitude (larch budmoth, *Zeiraphera diniana*; Baltensweiler and Fischlin, 1999): $N^* = 2.17$, $\alpha = 0.47$, $\beta = 1 \times 10^{-8}$ (arbitrarily set), $r(0) = 1.06$, $N(0) = 0.02$, $r_{max} = 5$. (G) Cycles of asymmetric shape (lemmings, *Lemmus lemmus*; Seldal et al., 1989: $N^* = 1.44$, $\alpha = 3.21$, $\beta = 1 \times 10^{-8}$ (arbitrarily set), $r(0) = 2.75$, $N(0) = 0.5$, $r_{max} = 5$.

specific mechanism. We prefer to think that the inertial properties are caused by maternal effects, but they could just as well be caused by the interaction with other species. This model is just a simple rendering of the second-order dynamics, with three parameters instead of the two in the logistic equation. This extra parameter dramatically increases the descriptive power of the model. In fact, this model is able to account for most observed data. It is not so powerful as to be trivial, however. That is, it is not so powerful as to be able to account for any arbitrary data set. For instance, this model will not be able to account for two distinct periods in the same data series. That such data series have not been observed suggests that our model strikes the right balance between simplicity and complexity; it's complex enough to handle existing data but is no more complex than it needs to be to do this.

In the last 30 years, a lot of attention has been devoted to analyzing population data series in terms of delayed density dependence. Extensive work by Royama (1992), Turchin (2003a), Stenseth (1999), and Berryman (1999), among others, has convincingly shown that we have to consider at least one lagged density dependence in addition to the immediate density dependence. That is, the two-dimensional model

$$N_{t+1} = N_t f(N_t, N_{t-1}) \tag{6.4}$$

has been shown to be the minimal description that is consistent with observed population abundance data series. This large body of work, although convincing on the question of the dimension in which the data series has to be embedded (it is two, rather than one), is unable to answer the question of the causes for the lag. It has this much in common with our implicit model, which is also silent on this matter.

A discrete analog of our model can be written as

$$\frac{N_{t+1}}{N_t} = \frac{N_t}{N_{t-1}} F\left(N_t, \frac{N_t}{N_{t-1}}\right). \tag{6.5}$$

This is equivalent to equation (6.4). The important difference is that by describing F with an expression with three parameters, r_{max}, α, and β, we have a great deal more power in the theory. For instance, just the presence of the term containing r_{max} turns locally unstable equilibria into limit cycles. Various oscillatory behaviors, both stable and unstable, are included. In the form of equation (6.4), eight parameters had to be introduced to faithfully reproduce the features of observed data series (Turchin, 2001). We believe that the reason we can achieve this with only three is the focus on acceleration, or the rate of change of the growth rate. Changing the response variable from the rate, to the acceleration reflects the inertial properties of population growth somewhat more naturally and thus allows us to achieve a general description with fewer parameters.

We are of the view that the maternal effect is the cause of inertia, but the fit to data and the particular parameterization of the implicit inertial model do not depend on this choice of the nature of the "hidden" dimension. Our arguments for the internal cause, based on observed periods of cycling, are independent of the particular form of the implicit inertial equation. This minimalistic form of equation (6.3) seems useful in itself as a way of describing inertial growth, irrespective of the cause.

6.3 Malthusian Invariancy

There is a well-known restatement of Newton's first law in terms of a thought experiment in which someone is placed inside a moving capsule (missile, train, boat, etc., depending on how the story is told). The restatement of Newton's first law is then the assertion that without looking out the window, or using other objects outside for comparison, the person will not be able to determine whether the capsule is at rest or moving uniformly. Only when the capsule accelerates (or decelerates) will the person be able to determine that something is happening. This is known as the *Galilean relativity principle*.

Indeed, such invariance, or symmetry, principles are fundamental to all of science. Whenever we have some significant quantity or relationship (e.g., length—the relationship between the end points of an object), we need to be able to say under what conditions this quantity or relationship will be unchanged and under what conditions it will change. If we can't do that, we can't claim to have a significant quantity at all. For example, in Newtonian physics, length is invariant with respect to velocity and with respect to position in space—the length of an object does not change if you change the object's velocity or location. That is, length is invariant under a certain class of transformations (known as *Galilean transformations*). (This example may seem trivial but it is not. As it turns out, length is *not* invariant with respect to Galilean transformation—the length of an object depends on its relative velocity.) Specifying invariances is necessary, it would seem, for the formulation of laws, for providing explanations, and for making predictions. After all, how can you predict what some significant ecological quantity (e.g., population abundance) will be in the future if you do not know anything about the conditions under which the quantity in question is invariant and under which it is not? To put it crudely, you don't know what something is until you also understand what it is not.

With the importance of invariance in mind, let us now try to formulate an ecological principle based on the Malthusian law that is analagous to the Galilean relativity principle mentioned above. This ecological principle will help demonstrate the connection between the ideas of inertial growth and ratio-dependent predation. So, in the spirit of the Newtonian thought experiment, imagine you are a cell in a laboratory cell culture or a rabbit in a population of rabbits (or even a human in a population of humans); how would you know whether the population that you belong to is constant in abundance, or whether its abundance is increasing (or decreasing) exponentially, by only looking around you, without conducting a census from the outside? We suggest that you would not know. A basic assumption of exponential growth is that there are no changes in the internal state of an average individual. We may, of course, determine the growth rate by counting, but we count "from the outside." We cannot internally sense the rate when we are members of the population being counted. The only events that we sense are individual, energetic-based changes. For example, if food becomes harder to obtain, or if you seem to encounter conspecifics more often, you would know that something is changing. This will correspond to changes in growth rate, not to the growth rate itself, and therefore, it is impossible to experience the rate directly (without counting from the outside).

Our proposal is that equations describing growth have to be based on the individual energetics that underlie population dynamics. The changes in individual energetics would not directly react to the population's growth rate. In the absence of data on individual quality, we may have to move to an implicit description, in terms of abundance and its rate of change. The corresponding second-order model will have to be invariant with respect to Malthusian transformation, just as the laws of physics are invariant with respect to the Galilean transformation. The meaning of this invariancy is that an individual in a population growing on an exponentially expanding resource will behave just as another individual in a population growing on a constant resource if the average energetic content of individuals is the same.

There is an important caveat to this statement. If the resource increases very fast, the population may not be able to reproduce as fast—there is an absolute limit for the maximum rate of reproduction. Therefore, the invariancy above is valid only approximately, when the growth rate is much below the maximum possible growth rate.

Imagine a situation in which two populations, prey and predator, equilibrate at some ratio, say, one predator per 100 prey. The absolute densities are equilibrated because of a limiting factor controlling the prey population. Now let us assume that this limiting factor (e.g., space, nutrients, or light) is available in abundance and the prey population grows for a period of time with the predator population growing along with it. If such growth continues, a ratio of abundance for the predator and prey (both increasing exponentially because there is no limit) will be established. The question is whether this ratio, with all other things being equal, is different from 1:100? In other words, the question is whether the predator population growth rate will react to the fact that the underlying resource is expanding rather than constant.

We have no doubt that such a reaction will take place with nonexponential expansion of the prey, because in this case the per capita supply of resources will change. The answer is not clear-cut in the case of purely exponential expansion. The Malthusian invariance principle suggests that the established ratio does not depend on the rate of expansion (or contraction, which might be easier to imagine). Change in the individual's quality is then the only absolute parameter "sensed" by predators. Evidence partially supporting this assumption comes from the comparison of steady-state densities at different trophic levels for lakes with varying nutrient concentrations. This evidence is far from compelling, but an approximate constancy of the ratios seems to be the case (Ginzburg and Akçakaya, 1992). We are not aware of any data that would allow us to truly test the exponentially expanding case.

In our daily lives, inertia of physical bodies is obvious. It would not be so obvious, however, if we lived in a very viscous liquid. Forces would then result in velocities, not accelerations, and the Aristotelian view would be sufficient. In physics, the gradient between inertial and friction forces in commonly described by the Reynolds number (Percell, 1977). At one extreme, when friction dominates, the Aristotelian view is sufficient; at the other extreme, inertia has to be taken into account. Limiting factors in population dynamics play the role in ecology that friction does in physics. They stop exponential growth, not unlike the way in which friction stops uniform motion. Whether or not ecology is more like physics in a viscous liquid, when the growth-rate–based traditional view is sufficient, is an open question. We argue that this limit is an oversimplification, that populations do exhibit inertial properties that are noticeable. Note that the inclusion of inertia is a generalization—it does not exclude the regular rate-based, first-order theories. They may still be widely applicable under a strong immediate density dependence, acting like friction in physics.

Even the Aristotelian limit in physics preserves a simple symmetry: only relative distances matter. Laws are said to be invariant with respect to translation. Ratio dependence within the traditional ecological theory is analogous to this: laws of interaction are invariant with respect to multiplying abundances of interacting species by the same constant. This, of course, is equivalent to the arithmetic "translation" in the logarithmic scale.

Space is particularly important here, and it is an ever-present limiting factor. It is therefore treated quite differently by the ratio-dependent approach. We separate biological *interactions* from biological *constraints*. Both are, of course, present in nature; the question is whether or not it is useful to treat them separately. We separate the two influences and assume just the predator-prey interaction to be ratio dependent. The limiting factors, like space, are relegated to a separate term in our equations, which stops otherwise infinite exponential growth of the interacting pair. According to the traditional line of thought, beginning with Lotka and Volterra, space is not to be thought of as a separate constraint but is included in the very basis of how interactions are described. Thus, traditional models not only disallow the joint exponential growth of predator and prey but also disallow even the thought of it. In other words, traditional models do not contain a parameter that, when set to zero or to infinity, will eliminate the limiting factor and allow for joint exponential growth. This *is* possible in the logistic model for a single species, but it is not possible in the model with species interactions. We believe that it is unreasonable to allow for the idea of Malthusian growth only in the case of single species. It seems to us that when limiting factors are introduced in the usual way in the Lotka–Volterra-like scheme, it is double counting—a pure scheme would have already reflected limited space, because the pure scheme is not invariant with respect to proportional scaling. The distinct treatment of biologically meaningful interactions and limiting factors is similar to the partitioning of forces in physics into true forces and the reactions of constraints. This partition is a useful, simplifying idea that may be helpful for constructing ecological models as well.

Our suggestion is to construct a theory of population dynamics that is invariant with respect to Malthusian growth. This invariance would lead to the ratio-dependent symmetry of interactions as a special consequence. As of now, we have more empirical justification for the consequence than for the larger view. It is more on the basis of beauty and simplicity than on direct evidence that we propose this view. There is certainly much more experimental and theoretical work to be done to decide whether a larger, Malthusian symmetry would be useful.

6.4 What Is and What Is Not Analogous

Now that we've spelled out the implicit inertial model of population growth, it will be useful to reconsider our analogy with Newtonian mechanics. We will show that, just like Newton, we are proposing a research program. Newton's research program was to identify physical forces; ours is to identify "ecological forces."

There are two ways of thinking about Newton's laws of motion. Recall that the first law tells us that bodies remain in a state of constant velocity unless acted upon by a net force. The second law tells us that forces act on bodies in such a way as to result in an acceleration that's proportional to the force. The gravitational law tells us about the gravitational force and its action on massive bodies.

On the first way of looking at these laws, the first law is redundant. After all, it seems to be a special case of the second law.

Set the force equal to zero in the second law and we see that the acceleration is zero, and hence the velocity must be constant. Why, then, do we need the first law? Some have suggested that Newton included the first law purely out of respect to Descartes and Galileo, who first postulated this law. On this view, the first law is a kind of redundant tribute to Newton's intellectual predessesors.

There is, however, another way of looking at these laws—a way in which there are no redundancies. On this second view, the first law is a statement about the default state of bodies—what happens, as it were, when nothing happens. The second law is seen as a *definition* of the concept of force: a force is defined to be a quantity that results in an acceleration. But this definition leaves completely open how many and what kinds of forces are possible. The two laws taken together, thus, open up a research program to identify the forces that exist in nature. The first step in this research program is taken by Newton with his gravitational law. This law identifies and describes the action of one such force: gravitational force. Of course, we know now that there are others. In fact, there are thought to be four fundamental forces: gravitational, electromagnetic, weak nuclear, and strong nuclear forces.

We prefer this second interpretation of Newton's laws. And in a similar spirit, we suggest that the first law of ecology is Malthus's law—the default state of a growing population. We *define* "ecological forces" to be quantities that act on growing (or declining) populations as second-order quantities—"ecological accelerations," as we've been calling them. We then identify one such force: the maternal effect, a force based on passing individual quality across generations. Also in the spirit of Newton, we do not rule out other forces (e.g., predator–prey interactions); what we propose is a useful framework for exploring ecological forces. The maternal effect, we claim, is one such force, and although we expect that there will be others, we think that a great deal can be explained simply by considering the maternal effect—just as a great deal can be explained in physics by simply considering one force, without resorting to the other three known physical forces.

We have been stressing the analogy between Newtonian laws and laws of population dynamics so much that the question of what is not analogous may be legitimately pondered. Let us take a moment to discuss what is not analogous, in order to clarify our position.

To start with, there is obviously nothing materially in common between how bodies move and how populations grow and interact. The equations are also different. There is no analogy to gravitational or electromagnetic force in ecology. The analogy helps inform the way we look at dynamics, and helps in generating new hypotheses in the research program that we propose to adopt.

As we've already discussed, the analogous program has been quite successful in physics. It has identified a number of forces, which can be captured in relatively simple equations, which in turn yield verifiable predictions. We have suggested an analogous research program, but we have not yet identified all the "ecological forces" or their mathematical descriptions. The three forces (read: causes of energetic changes for individuals or corresponding changes in growth rate) that we have addressed are as follows:

1. Energetics. We think that the balance of metabolism and consumption defines equilibrial abundances of organisms (Damuth allometry).
2. Maternal effect. We think that this is a very common mechanism of cyclicity. We think that it is a strong alternative to the traditional predator–prey account of the observed abundance cycles. The maternal effect account also fits well with the Calder allometry and produces the observed periods, while rejecting the unobserved periods.

3. Predator–prey interactions. We think that the better predator–prey model, consistent with our general framework, is ratio-dependent predation.

In all three cases, the forces of which we speak are formally analogous to physical forces in the sense that they result in changes in the growth rate (accelerations) rather than directly affecting the growth rate. We follow the Newtonian logic of approaching the problem, but that's *all* we do. In particular, we do not claim that ecological forces bear any resemblance to physical forces.

Seven

Practical Consequences

So far, we have argued for a different approach to modeling population growth—the inertial approach. The mathematical and theoretical differences between this and the traditional approach are clear. But is this just an academic debate? After all, if there are no *practical* consequences, why concern ourselves? We believe that there are practical consequences and that some of the limitations of traditional theoretical models may be due, at least in part, to adopting too narrow a theoretical framework. If the inertial model is the better model, it should fare better in applications. In this chapter, we outline a few of these applications. But first, it will be useful to reflect on the relationship between theoretical and applied ecology.

7.1 Theoretical and Applied Ecology

The beauty of general theoretical concepts is not what attracts students to ecology. Environmental concerns and the beauty and complexity of life remain the two main motives. Yet work in theoretical ecology—developing fruitful theoretical concepts and putting them to use in ecological theories—accounts for more than its fair share of the current and past research in ecology. Collections of the seminal papers in ecology contain 40–50% theoretical papers (depending on how you count them, of course). And this is with theoretical ecologists constituting less than 5% of all ecologists (Holland et al., 1992).

This dominance of theoretical output in ecology and conservation biology continues to this day. Both disciplines are in desperate need of theories that provide practical guidance. In particular, we need theories that help us answer such important questions as, How are we to protect various endangered species? Are we really in the midst of a mass extinction event? If we are, what can we do about it? How can industrial development and conservation coexist? The demand for answers to crucial questions such as these is the main reason for the existence of ecology and conservation biology as scientific disciplines. Indeed, almost all ecologists—theoretical and practical—owe their jobs to the importance of these practical matters concerning the environment.

In a social climate such as this, there is a great deal of pressure on ecologists to produce the goods—to produce theories that help us manage our environment. But good ecological theories are hard to come by; ecological theories are subject to natural selection no less than the biological populations they describe. There is no better selection pressure for a theory than the test of applicability. Not only will the theory need to produce results that match reality, but many models will simply crumble under the weight of their own complexity.

There are also problems of communication between the theoreticians and the more practically oriented ecologists. Practical ecologists often have trouble following the mathematical details of a theoretical argument. On the other hand, theoreticians often get so tied up in their mathematical models that they lose touch with the ecological system being studied. The result is that practical ecologists tend to treat complex theories with more than a little suspicion. As we shall show, we think that this suspicion is not without some justification.

The mutual understanding between theoreticians and mainstream ecologists has never been great. To start with, most older ecological theorists around today were not trained by ecologists. They came from physics and mathematics backgrounds and trained themselves in ecology by reading, by talking to biologists, and by working in biology and ecology departments. Because a typical biology department in a university rarely needs more than one theoretician, theoretical ecology is a minority occupation among ecologists—currently of the order of 5% of all ecologists. Theoretical topics are now taught to all ecologists, and new generations of theoreticians come almost exclusively from students trained within the profession. We hope this qualitative change will improve communication between theoreticians and the majority of ecologists. We say *hope* because the trend is counteracted to some extent by the general decline in the mathematical culture of biology students. It was never high, but it has continued to decline visibly in the United States over the last 20 years. European and Asian students fare somewhat better, but the trend is in the same direction. Either society at large does not require the degree of mathematical sophistication scientists would naturally like to see, or the trend may reverse as our children's children revolt against their parents.

7.2 Managing Inertial Populations

Although theoretical models in fisheries have been used for quite some time, it is only in the last 20 years or so that theoretical ecology has been widely put to practical use. This change of attitude toward theoretical ecology is mostly a result of the needs of conservation biology. It's fair to say, however, that although theoretical models are now fairly widely used, their usefulness in predicting the behavior of actual ecological systems is not all that one would hope for. There are, of course, many reasons for this, most notably the incredible complexity of the systems being modeled. But there may be another reason: the models may be fundamentally flawed.

Both in fisheries and in most conservation biology applications, the time scales of interest are much shorter than we have addressed by our suggested inertial model. Inertia works on the time scale of generations. If, as we've argued, cycles are minimally six generations long, then when a population overshoots or undershoots its equilibrial value, it will take at least three generations to return to that equilibrial value. Fish species—at least the ones of interest to humans—and many of the mammal and bird populations subject to conservation efforts have life spans that are often longer than one year and sometimes many years in length. The time scales of practical interest often have more to do with politics (e.g., presidential terms of office) than with biology. Most practically important theory has focused on the details of the age and stage structure of populations. The time periods of interest here are a few generation times of the species in question. The most common approaches are computer simulations of population dynamics that incorporate various sources of uncertainty. The results are formulated in terms of risk of decline or risk of extinction at time scales that are too short for incorporating effects of intergenerational inertia.

It is in the longer time scale of tens or hundreds of generations that our ideas will have practical impact. The main qualitative effect of the inertial view of population growth will be in the evaluation of the long-term effects of artificial mortality imposed on a natural population by harvesting, toxic effects, or any other source of human-caused mortality. In the traditional view, any additional mortality leads to the decline of the equilibrium population abundance. This is because birth and death rates are viewed as causes of equilibration, and with an increased death rate, the equilibrium abundance also declines. This interpretation remains valid for any source of mortality, including natural (e.g., predation) or artificial (e.g., harvesting). Thus, the response of the long-term equilibrium abundance to mortality is typically viewed as shown in figure 7.1A: abundance declines with mortality until mortality exceeds a critical level, and then the population is driven to extinction.

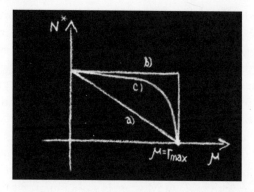

Figure 7.1. Response of equilibrium abundance, N^*, to additional mortality, μ. The three curves represent: traditional view (A), inertial view (B), and realistic expectation (C). When $\mu \approx r_{\max}$, equilibrium abundance is zero.

With the inertial view, the picture is quite different. The reason is that with additional mortality, and with constant resources, the population equilibrates, with the remaining individuals enjoying a higher average quality of life. This increase in quality will produce higher birth rates, which may compensate for the increase in death rate. Our view is based on a delayed (by one generation) density dependence. The picture of equilibrium response may look more like figure 7.1B. Equilibrium abundance may have little or even no response to mortality but then collapse when the mortality reaches the critical value. The critical value is the same in both cases: it is the maximum reproduction rate that, if exceeded by mortality, leads to extinction. The difference, however, is in the response to lower mortality levels. If the average quality does not respond in perfect proportion to mortality, or if the reproduction rate does not perfectly respond to quality, the picture will be more like that illustrated by figure 7.1C. It may not be as dramatic as a stepwise collapse (as in the extreme case), but it will be closer to a collapse response than to a gradual response. Therefore, if our view is correct, small rates of harvesting or small mortality rates in general may appear inconsequential. The compensatory response may fool us into believing that even higher rates of harvest are sustainable. We may then be surprised by a very rapid decline and even extinction of an exploited population.

Although the equilibrium may not noticeably respond to small amounts of imposed mortality, other characteristics of the population may. The so-called *resilience,* or rate of return to equilibrium after a deviation, will decline immediately after the imposed mortality, even if such mortality is small. In a fluctuating environment where average, rather than equilibrium, abundance is the more important concept, the average will decline even when the equilibrium does not decline. To what extent it shifts the response from gradual (figure 7.1A) to punctual (figure 7.1B) depends on the size and frequency of disturbances.

The argument above is applicable to natural mortality, as well. As we described in chapter 6, according to the ratio-dependence theory, efficient predators can consume prey to complete extinction. It is worth bearing in mind that, although naturally occurring pairs of predators and prey (and, indeed, whole trophic chains) have coevolved, it may well be that we only see the pairs that are able to coexist. This does not rule out past extinctions due to overefficient predation. It is likely that there were such extinctions in which both predator and prey disappeared. If such dual predator–prey extinctions occurred close to the bottom of a trophic chain, entire chains may have collapsed. On our view, dual extinction is a natural, possible outcome of predator–prey interactions. To deny this seems akin to believing that because the planets we see orbit the sun, every physical body in the solar system should also orbit the sun. But most bodies in the early solar system either fell into the sun (or, perhaps, what would become the sun), or fell into one of the nine planets (or their satellites), or were flung out into space. The outcome depends on the body's mass and its initial trajectory. Most trajectories are thus "selected against," and would-be satellites are selected according to the stability of their trajectory. It is an extremely unlikely combination of initial conditions that a body requires in order to occupy a stable orbit. Likewise, species that are present today are a small subsample of the ones that have existed. The ones we see do coexist, but a sensible theory has to allow for the possibility of extinction. Moreover, such a theory must give some guidance on the limits of values of parameters that, when exceeded, may lead to extinction.

The other side of the coin is the spectacularly unsuccessful attempts to control unwanted species, primarily agricultural pests, but also some wildlife. Deer populations in the United States, for instance, have become so abundant in some regions that various state governments are unable to control them by increasing hunting. At least part of the explanation for this failure may stem from factors similar to those discussed above. That is, if the response to mortality looks more like B than A in figure 7, the expectation of a reduction of abundance in response to mortality may be overestimated. Although the expectation might be close enough in the short term, it could be well off the mark in the long term—even with sustained mortality. If the inertial view is correct, simpleminded short-term policies that fail to take into account the quality of individuals are destined to fail.

It is worth stressing that this inertial, or two-dimensional view of equilibrium is not a rejection but a generalization of the traditional rate-based view. It is possible that, with very low inertia (because of a small effect of individual quality on the rate of reproduction) and with strong direct limitations to growth (unrelated to resource consumption), the traditional rate-based equilibration will work well. We certainly do not rule that out. Indeed, our view allows this as a special case, but our inertial view also opens the door to other possibilities that may have significant implications for management strategies—possibilities that cannot even be entertained on the traditional view.

7.3 Rates of Evolution

Since its first formulation, the theory of evolution has been deeply rooted in the Malthusian model of population dynamics. Adaptations are thought of as changes in the rate of reproduction. Natural selection acts by replacing the less fit (lower rate of reproduction) with the more fit mutant (higher rate of reproduction). The inertial growth model adds a seemingly small, but potentially significant, twist to this view. A mutant achieving a higher reproductive rate at the same energetic cost would win the competition. However, because in our model reproduction rates explicitly depend on energetic investment, there is another option. A more energetically efficient mutant will win the competition, and this process can occur much faster than in the first case. Because resource investment in reproduction and maintenance is taking place at the energetic level, the underlying reproductive changes of an advantageous mutation will cause a superexponential or accelerated rate of replacement by a mutant. We suspect that the rate of evolutionary change can be much higher than previously thought if the inertial population growth model is placed at its foundation. This is not an issue that we have fully explored, and our suggestions here are rather speculative.

Another aspect of evolutionary applications involves our proposed oscillatory image of a single species population consuming a constant flow of the resource. If this view is correct, one possible strategy for the evolution of species interactions is *resonance avoidance*. Because the periods of oscillations for a species, according to our view, are crudely a function of the species' generation-time eigenvalues, interacting pairs have to have sufficiently disparate generation times in order to avoid high-amplitude resonance and thus an increased chance of extinction. In fact, as Elton (2001) pointed out, predators commonly have a much longer generation time than do their prey. Parasites have a much shorter generation time than do their hosts. The usual explanations are based on body sizes as causes and generation times as effects for these structures. The inertial view, however, suggests that it may be the other way around. It may well be that interactions between species with similar generation times are dynamically unstable and that resonance avoidance is the reason why stable trophic chains consist of groups of species with very different generation times. For example, consider the trophic chain of phytoplankton (generation time in days), zooplankton (generation time in weeks), forage fish (annual or biannual), and predatory fish (multiyear generation).

Another possible evolutionary application of our account of population growth could involve capitalizing on inequality (5.1), controlling dual extinction versus coexistence of predator and prey. If the predator is too efficient, it eliminates the prey and then dies, as in the Gause experiment. This is certainly not the usual individual selection mechanism. We submit that it is also not the usual group-selection or species-selection mechanism. A good name for it may be *ecological exclusion*; it is a systemic or structural elimination, not a kind of fitness competition in either the individual or the group sense.

Such elimination may have happened in the past, and today's predator–prey pairs satisfy the coexistence inequality by not having predators that are too efficient. A similar argument was repeatedly raised in epidemiology. Moderation in virulence and apparently "prudent" predation in the sense introduced by Slobodkin (1980) may need a systemic explanation. A well-known metaphor by Hutchinson sets ecology as a theater and evolution as a play. To continue with this metaphor, it may be that not just actors compete and are eliminated, but whole plays may be excluded with all their actors, when characteristics of the pair of species cross the coexistence line. We would thus have two quite different methods of elimination. This possibility has been tentatively suggested by Dunbar (1971), but given that traditional prey-dependent theory does not allow dual extinction as an outcome, the suggestion was not pursued. If predator dependence symbolized by ratio dependence is a better idealization, this evolutionary mechanism has to be given serious consideration. The whole body of the spectacularly successful evolutionary theory has Malthusian growth in its foundation. An incremental progress in understanding population dynamics a step away from this simplest law of growth has to fuel substantial progress in understanding evolution as well.

There are, no doubt, many more questions about evolution that could be asked and possibly answered differently if we adopted the inertial view of population growth. It is beyond the scope of this book to explore all such issues. Our purpose here is simply to highlight the fact that the inertial model does have serious evolutionary implications, and to briefly sketch what some of these might be.

7.4 Risk Analysis

Populations grow when their abundance is low in relation to the resources used for reproduction; they decline when their abundance is high in relation to resources. Somewhere between the low and the high lies the equilibrium value of abundance. If this value is stable, and if a population abundance approaches this value in an appropriate way, the abundance will stay at the equilibrium value. Populations in nature are, of course, rarely in perfect equilibrium. Unpredictable variability of environmental characteristics shifts the abundance of populations up and down. The effect of this unpredictable variation is often pronounced and may have consequences for the value of average abundance (which need not be equal to the equilibrial abundance). The rate of return to equilibrium, for instance, may have a strong effect on the average abundance, even when the equilibrium itself is unchanged.

Such variability, or environmental fluctuations, has led to a reformulation of one of the key questions asked by applied ecologists. Instead of predicting exact values for future population abundance, ecologists are asking for only an estimate of the risk of population decline or extinction. Methods for assessing such risks have been developed over the last 20 years. They are usually implemented using software, because the required Monte Carlo simulations need significant computing power. The time scale of the ecological risk assessment is often quite short—at least if thought of in terms of generation times of the species in question. The reasons for such short time scales are the urgency of the issues under consideration and the way error accumulates to make longer term assessments unreliable. The mathematical models used for ecological risk assessments are typically based on age- or stage-structured demographic models. They describe the growth of populations with much shorter time units than is generally found for generation times. Thus, the inertial growth view, which is applicable in a much coarser time scale, has not been incorporated into risk analyses. On the other hand, long-term population data series accumulated at Imperial College in London contain a significant number of long-term population waves (NERC, 1999). This, in turn, suggests the need for an inertial description at these longer time scales. It appears that with an increase in our understanding of basic laws of population dynamics, the inertial view may be incorporated into practical risk assessments aimed at longer time scales.

7.5 The Moral

Why do most attempts to control natural populations fail? From fishery collapses to pest explosions and dismal attempts to control various wildlife species, the news has been generally negative. Under the U.S. Endangered Species Act, more than 1,200 species have been listed as threatened or endangered. Of those, only nine have been delisted (deemed to have recovered). With all the effort put into conservation management and, in particular, the protection and regeneration of threatened species, why such a poor recovery rate? The blame is commonly placed on shortsighted government policies, management errors, the poor data on which predictive models are built, and the unpredictability of environmental changes. We believe that a portion of the blame may also be attributed to the failure of standard ecological theory. This theory is the very basis of the predictive models underpinning population forecasts and on which management and conservation efforts are planned. It may well be that dependence on traditional population dynamics is one of the reasons for the failure of many environmental management policies.

Viewing populations as inertial may lead to a shift in focus in the design and implementation of practical strategies for harvested or controlled population management. Using various sources of mortality as controls is bound to lead to populations substantially overshooting the desired abundance both in amplitude and in duration. Thus, the inertial model may be viewed not simply as an abstract exercise, but as a potential basis for the revision of population management. However, the favorite time horizon of governments, the "5-year plan," may not be an appropriate time scale for sensible management of relatively long-lived species.

It is instructive to learn that until about 20 years ago fisheries management was commonly based on the assumption that harvest does not affect recruitment. This is certainly not true, but high fecundities of fish disguised this fact. High fecundities created an impression of practically unlimited "surplus production." Enforcing a quota for fishing as a method of saving the overfished stock, spraying insecticides to control pest growth, and imposing additional mortality on an overgrown deer population may not be the best ways of achieving the desired results. A more sophisticated approach, based on monitoring and dynamic feedback, might be required. This is, of course, old news in other disciplines such as engineering and physics, which have long traditions of controlling inertial objects. But in ecology, such monitoring and dynamic feedback are new. Moreover, the need for monitoring and feedback in ecology is emphasized all the more by the inertial model. Certainly, because the inertial character of population growth has not yet been recognized, the consequences for management have not been fully appreciated.

Consider an inexperienced boat captain who does not appreciate the inertial nature of the vessel she has at her command. Such a captain, when coming into dock, would decelerate the boat only at the very last moment. Of course, such a docking strategy would result in a violent impact between boat and dock. Our current ecological management strategies may well be like this. But in the ecological case, the consequences are even more serious—they can be *extinctions*. An experienced boat captain, on the other hand, knows to start the deceleration well before the boat gets to the dock, and learns that a successful docking requires a subtle series of decelerations and sometimes even accelerations in response to the boat's movements. We suggest that theoretical and applied ecologists alike need to appreciate the inertial character of populations and learn the subtle dynamic-feedback strategies required for successful management.

Eight

Shadows on the Wall

It can be very difficult to decide between two competing theories. It is rarely a simple matter of appealing to evidence. For one thing, typically, neither theory conforms perfectly with the evidence—usually some auxiliary hypotheses are needed for this. But more important, often *both* theories can be made to agree with the evidence, by a suitable choice of auxiliary hypotheses. The issue, then, is not which theory best accords with the data, but which does so in the simplest or least ad hoc way.

Appeals to the notion of simplicity in science are rife with controversy. For example, there is the nontrivial issue of spelling out what simplicity amounts to. In this final chapter, we look at two ways to get a theory to conform to complicated data sets. The first is the method of increasing the dimension of the model; the second method involves increasing the number of parameters of the model. The inertial theory of population growth is an example of the first strategy—it suggests a *second*-order model of population growth rather than the traditional *first*-order model. One of the virtues of the inertial theory is that it is simple in the sense that it has few independent parameters. We now explore these issues in more detail, and we mount a case for at least sometimes preferring the strategy of increasing the dimension of the theory. Finally, we discuss another way in which our theory is simpler: it makes possible a rather elegant image of ecological interactions.

8.1 Plato's Cave

Plato thought that all we see around us are mere shadows of the "real" world. Unlike the imperfect world we seem to inhabit, the real world is perfect. He used a now famous metaphor to illustrate his view. He suggested that it's as though we are inhabitants of a cave and all we are allowed to see are flickering distorted shadows on the irregular cave wall. From this we must try to infer what the real world that casts these shadows is like. Putting aside Plato's contentious metaphysical views, the cave metaphor is a wonderful statement of the scientist's plight. We are faced with data that are often unreliable, biased in various ways, and often permit many interpretations—just like the shadows on the cave wall. But from these imperfect data we must construct a picture of reality that is much more than the sum of all the data. This picture should be free of the imperfections we see in our data and it should tell us how things *really* are.

Imagine, for a moment, that you're in Plato's cave, looking at the flickering shadows on the wall. How many dimensions will you need to explain the fact that some shadows seem to pass through one another while others seem to bounce off one another? You could doggedly stick to two dimensions, because your data are two-dimensional. You might insist that yours is a theory of shadows and that there are many different kinds of shadows: some pass through one another whereas others do not. Alternatively, you could make an abstract leap and suggest that the objects that the theory is about are in fact three-dimensional physical objects inhabiting a three-dimensional world, of which you are merely seeing the two-dimensional projection—and an imperfect projection, at that. Now, the explanation of why some shadows seem to pass through one another is straightforward: all solid physical objects bounce off one another, but what you see on the cave wall as shadows passing through one another are just physical objects passing by one another at different distances from the cave wall. Although the dimensionality of theories was not what Plato had in mind with his analogy, nevertheless, it's a wonderful illustration of the point that sometimes the best explanation of the data is at a higher dimension.

The history of science has many examples of increasing dimensions to account for, and explain, the data. As we've repeatedly stressed, Newton and Galileo suggested the move from first-order equations of motion to second-order equations of motion, thus increasing the dimension of the relevant theory by one. Einstein and Minkowski suggested that the appropriate model of space is in fact four-dimensional—where time is a spacelike dimension and the four together are thought of as a space-time manifold. In modern quantum mechanics, the dimensionality of the relevant Hilbert spaces employed is infinite. The suggestion that we advocate for population ecology is a very modest one. We are suggesting the move from a first-order model to a second-order one. But the question remains, when are such increases in dimensions justified and when are they not? And, more pertinent to the point of the present discussion, is the move from the traditional Lotka–Volterra worldview to the inertial theory of population growth legitimate or not?

We don't believe that there is a general answer to the former question. We must consider each case of increasing the dimension of the theory on its own merits. But it is clear from the many successful applications of increasing the dimension that it is at least sometimes justified. As for the latter question, we agree that, all other things being equal, we should prefer the simplicity of lower dimensional theories. The question, then, is whether the inertial theory is the simplest theory that can account for what we see. This issue clearly requires further investigation, but we hope that we have said enough in the course of this book to show that the inertial theory of population growth is at least a viable alterantive to its more traditional rival.

8.2 Evidence and Aesthetics

The scientific method, as anyone will tell you, compels us to look impartially at the evidence and deduce our scientific theories from that evidence. Although this view is widely held, it is, nonetheless, a fiction. In fact, it is not even clear that there is any single method by which all science proceeds. In any case, the scientific method (if there is one) is much more subtle and much more interesting than this simplistic picture suggests. It is instructive, however, to see where this picture goes wrong.

The first problem with this simple view is that the evidence is often ambiguous or does not wholly support or wholly fail to support the hypothesis in question. Think of the standards of evidence in legal proceedings. The best and clearest evidence supports the hypothesis *beyond reasonable doubt*, whereas weaker evidence might be merely *clear and convincing*. Still weaker evidence is when the hypothesis is supported merely on the *preponderance of evidence*.

If we admit various standards of proof, as it seems we must, then science is not simply a matter of looking at the data and producing a theory that fits the data—there are degrees of fit. Moreover, even this order of events can be questioned. Very often we have a theory *first* and the theory suggests what data we ought to seek and how we should interpret it. We are not suggesting that you see what your theory tells you to see, just that empirical evidence is very often *theory laden*. For example, when a doctor "sees" a chest infection in an x-ray, she does so only via a certain amount of theory about how x-rays work and about the relative absorption of x-radiation by infected lung tissue.

The next problem with the statement of the scientific method presented above is that it suggests that theories arise from evidence alone—it ignores the role of aesthetics in science. It may surprise some to learn of the deep and important role that aesthetic considerations play in science, but many scientists have waxed lyrical on this. For example, the scientist Henri Poincaré (1854–1912) once said: "A scientist worthy of the name, above all a mathematician, experiences in his work the same impression as an artist; his pleasure is as great and of the same nature" (quoted in Bell, 1953, p. 526). The physicist Richard Feynman (1918–1988) said: "To those who do not know mathematics it is difficult to get across a real feeling as to the beauty, the deepest beauty, of nature" (Feynman, 1965, p. 58).

There is a tendency, when reading passages such as these, to simply dismiss them as signs of scientists who have gotten a bit sentimental and somewhat mystical. Although the mood in these passages is decidedly sentimental, the message is very serious and essentially correct: there is most definitely an important role for aesthetic considerations—especially simplicity—in standard scientific methodology. Indeed, one of the best examples of beauty in science can be seen in Darwin's simple yet powerful idea of natural selection.

To understand the role of aesthetics in theory choice, we need to see that empirical evidence is not enough to determine a unique theory. Indeed, it is not too difficult to show that there are many (arguably infinitely many) theories that conform to any set of observations. For example, consider the following theory, attributed to William Gosse (1810–1888), that competes with standard evolutionary and geological theory: the earth was created by God about 4,000 years ago with all the fossil records in place. No amount of evidence can help us discriminate between these two theories. We take it, however, that Gosse's theory should be rejected, and evolutionary and geological theory should be accepted. Our point is simply that the rejection of Gosse's theory cannot be based on purely empirical grounds because it, too, agrees with the evidence. Typically, such decisions are made on aesthetic grounds. Standard theory is more natural, simpler, less ad hoc, more unified. Again, this is not meant to be a skeptical conclusion. We *can* justify our belief in standard evolutionary and geological theories over Gosse's theory; it's just that this justification must invoke aesthetic considerations—it cannot rest on empirical matters alone.

There is one last concern with the simple statement of the scientific method at the opening of this section. The concern is that the simple view suggests that we can tell when a theory conflicts with the data, and that when a theory does conflict with the data, the theory should be rejected. This, too, is highly questionable. It is very common for scientists to dismiss certain pieces of data because such data conflict with accepted theory. As Robert MacArthur (1930–1972) rather nicely put it:

> Scientists are perennially aware that it is best not to trust theory until it is confirmed by evidence. It is equally true, as Eddington pointed out, that it is best not to put too much faith in facts until they have been confirmed by theory. (MacArthur, 1972, p. 253)

Statisticians routinely, and quite rightly, reject as *outliers* data points that do not conform to the relevant theory; physicists attribute certain anomalies to experimental error and then ignore them. For example, Einstein, when asked whether he would have been concerned if the results of the crucial 1919 Eddington eclipse experiment came in against general relativity, he replied that if the experiment contradicted the theory, the experiment would have been in error; the theory was too beautiful to be wrong.

In some cases, recalcitrant data are not ignored, but in order to save the theory, "epicycles" are added to the theory so that it may account for the data. The most famous case of this was the Ptolemaic model of the solar system. The simple, Earth-centered model, with the planets, the moon, and the sun circling Earth, did not fit well with observations. Although the sun's and the moon's orbits were well described by the model, the planets' orbits were problematic—they exhibited *retrograde motion*, for instance. (This is the apparent, occasional backward motion of the planets, as viewed from our earthly perspective.) But instead of giving up the Earth-centered model, it was merely modified: the planets were assumed to be orbiting points that were, in turn, circling Earth (these secondary cycles are the so-called *epicycles*). When *this* epicycle model conflicted with the data, further epicycles were added.

Most current predator–prey theories of population cycles are, in our view, another example of this Ptolemaic way of thinking (Ginzburg and Jensen, forthcoming). In the predator–prey theory, we have a sequence of models of increasing complexity, and we even have parameters fitted to the models that produce behaviors strongly resembling major aspects of the historical data. Our advice is not to bet on these models. The number of parameters is in a way a measure of how much massaging one had to do to fit the theoretical construct to the evidence—how much trial and error went into producing the model. Good theories work well *from the start*; they are simple and general and have few parameters. They account for the data in a natural way. Therefore, one rather crude but objective measure of the simplicity of a model is the number of parameters it requires.

8.3 Overfitting

Let us return to our favorite planetary example. Assume that we are living in pre-Newtonian times and we already know from Kepler's discovery that the orbits of planets are elliptical, though we do not yet know any more than that. We can, however, develop a taste for the future theory. Consider the question, How many parameters do we have to specify to describe an ellipse? The answer is five. We have to know the locations of two foci on a plane (each focus requires two numbers to specify its location), plus we need the length of the string attached to the two nails to draw a specific ellipse (see figure 2.6). If the theory to be evaluated has more than five parameters, we would not be too impressed by it. The theory may be able to describe our ellipses, but it will have a lot of freedom to produce many other curves. Because we do not observe these other curves, we judge such a theory as "overfitted."

One of the many virtues of Newtonian gravitational theory is that it has exactly five parameters. These parameters include two initial conditions for the relative location and speed of the planet with respect to the sun, plus the mass of one of the two bodies. The theory produces not just the ellipse for one planet but, as soon as five values are correctly specified, it also produces ellipses for all the planets. In addition, without extra parameters, the theory produces various other trajectories of bodies falling into the sun, into planets, flying away, and so forth. Indeed, if initial conditions plus the masses are not right to make a given body a planet (i.e., able to maintain a stable, periodic trajectory), the body in question will have one of these other trajectories. The theory is beautiful because it is minimal.

Now, back to population-abundance cycles. The simplest nonmechanistic description of a cycle is a sine wave. It requires two parameters: period and amplitude. If we wish to pay attention to cycle asymmetry (slower increasing than decreasing), we may need another parameter. If we have data for two interacting species, assuming that the period is the same, we need another amplitude for the second species and the so-called phase shift, or a time-lag value between the two waves. So far, we have five parameters. If the theory of predator–prey interaction (Turchin and Batzil, 2001; King and Schaffer, 2001; Hanski et al., 2001) had any more than five parameters, we should be inclined to think that it is overfitted—it's overly complex for the problem at hand. So, in particular, some of the predator–prey theories in the current literature, with more than 10 parameters, offend our aesthetic sensabilities. Theories such as these eventually collapse under their own parametric weight.

On the other hand, our very simple model of cycles based on maternal effects uses two parameter values, one of which (the strength of the maternal effect) is not known. It explains the main feature of all the cycles: their periods. That predator–prey interactions can cause cycles of abundance is undeniable. Whether they actually do is another question. We believe that we have accumulated substantial circumstantial evidence toward the inertial view. This includes an explanation of the Calder allometry, the gap in the observed periods measured in generations, and the quantitative fit of periods and shapes of oscillations for all cyclic species to a simple, general (and not overparameterized) model. All of our evidence combined does not constitute a proof, but again, it makes our view a serious rival to the traditional explanation.

8.4 A Simplified Picture of Population Ecology

We have argued for a second-order inertial model of population growth. This model is able to accommodate more complex population abundance behavior (e.g., population cycles) without having to appeal to population interactions (e.g., predator–prey interactions). The reason for this second-order behavior, we argue, is the maternal effect. This effect involves the inheritance of quality from mother to daughter. This, in turn, means that population abundance, at a given time, is dependent on both the current environment and the environment of the preceding generation. This time lag lies at the heart of the second-order model we propose.

We have also suggested that the inertial model is simpler than its traditional competitors, and this, of course, is a good thing. Very little population data are available, and what data are available do not justify complex theories. In most cases, exponential growth, with a fluctuating growth rate (random walk of abundance in the logarithmic scale), is as complex a model as can be justified based on the evidence. Theories in ecology, particularly models of interacting species, have systematically strayed from the evidence. More and more complex phenomena have been "discovered" in systems of differential equations in which no single parameter is reliably measured. Even a unit of time, the dt in dN/dt, has not been clear in most cases.

We have attempted to make a fresh start, beginning again with Malthusian growth and suggesting that, because of the maternal effect, population growth is an inertial process. In any case, the data can be accounted for by a simple, single-species, inertial model acting on the time scale of generations. We believe that the burden of proof for inclusion of species interaction has to be shifted to the theoreticians building complex models. They have to point to observations that would require more complex theories. The model of an inertial single species connected to a single resource is able to describe most of the patterns we see.

With the inertial view in physics, much of the complexity of the solar system can be described by a two-body interaction. In fact, the problem can be reduced to a one-body problem, where the body in question is simply thought to be in the gravitational field of another. Planets certainly interact with each other and their moons are affected by the sun as well. It turns out, however, that these effects are secondary and, for many purposes, can be ignored. The nested set of ellipses—planets orbiting the sun and moons orbiting planets—describes planets and their satellites quite well, and it does so without invoking higher order interactions.

Populations may "orbit" their respective resources, not unlike planets orbiting the sun and moons orbiting their planets (hence the illustration on the cover of this book). Population growth responds to changes in resources with a varying degree of inertia, sometimes undergoing noticeable cycles of over- and undershooting the average level. Various limiting factors act as friction. In most cases, these limiting factors are so prevalent that oscillations are absent or hard to see. Environmental fluctuations make population orbits look noisy, so careful analysis is needed to reveal hidden waves that are multiple generations long. Although this picture looks complex, it is much simpler than the prevalent view of insurmountable complexity when all species interact with all other species (see figure 8.1). The nested structure of orbits stresses approximate dependence of population growth only on the population's own resources. By including inertia and making the single-species view a bit more complex, we simplify the big picture. It is as if herbivores orbit the plants and carnivores respond only to herbivores, with no regard to the plants, just like the moon orbits Earth irrespective of whether Earth is orbiting the sun.

Another "big picture" metaphor in ecology is the *Eltonian pyramid*, where each trophic level is seen as a level in a resource pyramid. According to this view, biomass declines as we go up the trophic levels. Plants are at the foundation and have the greatest biomass. Above them in the pyramid are herbivores, and above them carnivores. In some ways this view is like our own: the herbivores depend only on the plants and the carnivores depend (directly) only on the herbivores. But this seductive picture is too simple. Some herbivores, such as elephants, lack predators, whereas others, such as rabbits, have many predators. Our picture has no such shortcomings. Planets can have zero, one, or many moons, just as different species can have zero, one, or many predators.

This nested "planetary" structure of ecosystems is the image that we propose to replace both the oversimplified Eltonian pyramid and the overly complex spaghettilike diagrams of hundreds of species interacting with each other. Of course, we do not deny that such complex interactions occur. Populations, like planets, interact with almost everything around them. The question is, How much do these interactions need to be acknowledged by our respective theories of populations and planets? The answer, in the case of planets, is that all but the gravitational forces due to the effect of the planet's sun can be ignored (at least for most purposes). Our answer for ecology is much the same. We may be able to make a great deal of progress by considering only

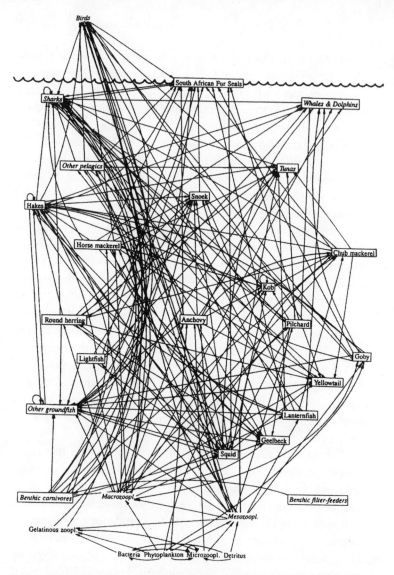

Figure 8.1. The Benguela ecosystem, South Africa, represented as a food web diagram (Abrams et al., 1996, p. 387), reprinted with permission from Chapman and Hall.

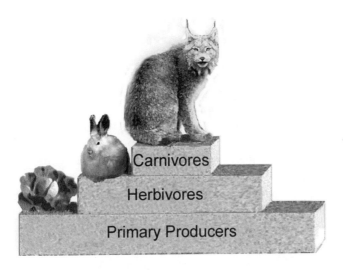

Figure 8.2. Eltonian trophic pyramid.

the population in question and its resources. And this simplification is made possible by considering the inertial properties of population growth.

Here is a summary of what we have proposed in this book:

- In ecology, the law of Malthusian growth plays a role analogous to Newton's first law in physics. Malthusian growth describes the background of ecological events that are interactions, both internal and external.
- Population ecology has to focus on the deviations from Malthusian growth rather than on deviations from constant abundance. Therefore, ecological models have to include inertia, or second-order effects.
- Even though, at equilibrium, net growth rates of population abundances have to be zero, it is individual quality that is the cause of the equilibrium.

- Maternal effects cause inertia of population growth on the generation time scale. Resulting cycles are at least six generations long and are consistent with observed periods for cycling populations. It is the absence of observed periods in the 2–6 generations range that is the most convincing evidence for our view. The relative role of species' interactions and internal inertia may be that of disturbance and every species' own eigenperiod.
- Ratio dependence, in describing predator–prey interactions, may be a special case of a larger invariance of interactions with respect to Malthusian transformation. The former may be the simplest symmetry of ecological models, with more expected to be discovered in the future.
- The inertial view of population growth, in its implicit form (as a single, second-order equation), is a simple generalization, not a rejection, of the traditional view. It achieves substantial flexibility with a minimum of parameters, and naturally incorporates a biologically meaningful upper limit to the growth rate.
- The practical consequences of the inertial view for applied ecology relate to longer time scales than commonly considered, because the effects have multiple-generation time scales. Neglecting to account for inertia may be the reason for systematic errors in management policies for various natural populations, particularly for species with longer generation times.
- If our view of population dynamics is generalized to include inertial effects, evolutionary theory may have an additional impetus to follow. Evolutionary theory certainly has enjoyed spectacular success with the Malthusian-growth model at its foundation. In view of our planetary analogy, we suggest that a form of systemic "ecological exclusion" needs to be considered, and this may lead to new developments in evolutionary theory.

- Our proposal is a *research program*. It changes some of the questions typically raised in population ecology. A few of the consequences of the program are found in our suggested explanations of some of the body-size allometries and periods of cycling. We expect more consequences to be discovered if our program is pursued.

Appendix A: Notes and Further Reading

Frontmatter and Preface

The quotation from Robert MacArthur is from his last paper (1972), which he wrote from his hospital bed. In this paper he repeatedly makes comparisons between ecology and physics. The dialogue is from the movie *Il Postino* (released by Miramax in 1995) and is based on the novel *Il Postino di Neruda* by Antonio Skarmeta (published by Garzanti, mid-1980s).

Chapter 1

The analogy between Newton's first law and exponential growth was suggested independently by Clarke (1971) and Ginzburg (1972). The better known references are Ginzburg (1986) and the discussion in Ginzburg (1992). Although the analogy had been accepted by other ecologists (Berryman, 1999; Turchin, 2003a), there are substantial differences between various proposals describing the deviation from exponential growth. Our proposal stresses the internal causes, and is described in chapters 3 and 4. Berryman and Turchin take a more traditional stance and cite species interaction as the cause of the deviation. See Kingsland (1985) for a good history of ecology.

A very nice popular account of the physics we discuss in this chapter and elsewhere can be found in Feynman (1965). See

Feynman et al. (1963) for a more rigorous and thorough introduction to basic physical theory.

Chapter 2

Interest in allometric relationships, including ecologically related ones, is growing very fast. Many books have been published recently on this topic (Calder, 1984; Reiss, 1989; Harvey and Pagel, 1991; Charnov, 1993; Brown, 1995; Brown et al., 2000). Of these Calder (1984; 2nd ed., 1996) is, in our judgment, still the best. William Calder died in 2002, in the very month he had scheduled a seminar at Stony Brook. We thus missed the chance to meet him. His allometry for the period of cycling species offers some of the strongest support for the inertial view of cycling. As a small tribute to William Calder, we suggest that the allometry in question should bear his name and we include it in the list of the basic statistical observations, which are for ecology what Kepler's laws are for classical celestial mechanics. Ecological allometries are a part of a new field of macroecology (Brown, 1995). A review by Lawton (1999) stresses the great potential of this field while expressing some skepticism about our ability to understand the species interactions that underpin it.

For good general discussions of laws of nature, see Armstrong (1983), Newton-Smith (2000), Chalmers (1999, esp. chap. 14), van Fraassen (1989, chap. 2), and Smart (1968, esp. chaps. 3 and 4). Quine's influential views on the philosophy of science can be found throughout much of his work. Good places to start, however, are Quine and Ullian (1978) and Quine (1981, 1995). His discussion of limit myths can be found in Quine (1960, esp. pp. 248–251). Not everyone agrees that there are laws of nature. See Cartwright (1983) and van Fraassen (1980) for some skepticism in this regard, and see Cartwright (1999) for a very interesting discussion of the alleged unity of science. Debate about whether ecology has laws can be found in many places, but see

Appendix A: Notes and Further Reading

Frontmatter and Preface

The quotation from Robert MacArthur is from his last paper (1972), which he wrote from his hospital bed. In this paper he repeatedly makes comparisons between ecology and physics. The dialogue is from the movie *Il Postino* (released by Miramax in 1995) and is based on the novel *Il Postino di Neruda* by Antonio Skarmeta (published by Garzanti, mid-1980s).

Chapter 1

The analogy between Newton's first law and exponential growth was suggested independently by Clarke (1971) and Ginzburg (1972). The better known references are Ginzburg (1986) and the discussion in Ginzburg (1992). Although the analogy had been accepted by other ecologists (Berryman, 1999; Turchin, 2003a), there are substantial differences between various proposals describing the deviation from exponential growth. Our proposal stresses the internal causes, and is described in chapters 3 and 4. Berryman and Turchin take a more traditional stance and cite species interaction as the cause of the deviation. See Kingsland (1985) for a good history of ecology.

A very nice popular account of the physics we discuss in this chapter and elsewhere can be found in Feynman (1965). See

Feynman et al. (1963) for a more rigorous and thorough introduction to basic physical theory.

Chapter 2

Interest in allometric relationships, including ecologically related ones, is growing very fast. Many books have been published recently on this topic (Calder, 1984; Reiss, 1989; Harvey and Pagel, 1991; Charnov, 1993; Brown, 1995; Brown et al., 2000). Of these Calder (1984; 2nd ed., 1996) is, in our judgment, still the best. William Calder died in 2002, in the very month he had scheduled a seminar at Stony Brook. We thus missed the chance to meet him. His allometry for the period of cycling species offers some of the strongest support for the inertial view of cycling. As a small tribute to William Calder, we suggest that the allometry in question should bear his name and we include it in the list of the basic statistical observations, which are for ecology what Kepler's laws are for classical celestial mechanics. Ecological allometries are a part of a new field of macroecology (Brown, 1995). A review by Lawton (1999) stresses the great potential of this field while expressing some skepticism about our ability to understand the species interactions that underpin it.

For good general discussions of laws of nature, see Armstrong (1983), Newton-Smith (2000), Chalmers (1999, esp. chap. 14), van Fraassen (1989, chap. 2), and Smart (1968, esp. chaps. 3 and 4). Quine's influential views on the philosophy of science can be found throughout much of his work. Good places to start, however, are Quine and Ullian (1978) and Quine (1981, 1995). His discussion of limit myths can be found in Quine (1960, esp. pp. 248–251). Not everyone agrees that there are laws of nature. See Cartwright (1983) and van Fraassen (1980) for some skepticism in this regard, and see Cartwright (1999) for a very interesting discussion of the alleged unity of science. Debate about whether ecology has laws can be found in many places, but see

Lawton (1999), Quenette and Gerard (1993), Slobodkin (2003), Turchin (2001), and Colyvan and Ginzburg (2003a, 2000b). The issue of what scientific explanation amounts to is a large and complex one. Hempel's classic (1965) is a good starting point. For the record, we are inclined toward the unification account of explanation. Kitcher (1981) gives a nice outline of this philosophical theory of explanation.

Chapter 3

A very nice and accessible account of Galileo's work can be found in the recent bestseller by Dava Sobel, *Galileo's Daughter* (1999). Papers on accelerated death by Akçakaya et al. (1988) and Ginzburg et al. (1988) analyze the Slobodkin experiment. The Damuth (1981) allometry has been questioned on various grounds, but it is becoming more accepted with time. Our argument, relating this allometry explicitly to Kleiber's (1975) rule, is new and, in our judgment, strongly suggests that energetic processes underpin a great deal of population dynamics.

Strictly speaking, the Damuth allometry is observed only when the amount of energy resources each population has available per unit time, S, is independent of body size across species. That is, species in general should not differ in their potential for extracting energy from the environment solely as a function of their size. As noted in chapter 2, the observed exponent of $-3/4$ means just this (Damuth, 1981, 1987, 1991); some have therefore referred to the Damuth allometry as the "energetic equivalence rule" (Nee et al., 1991). This assumes, of course, ratio-dependent consumption. This assumption is accepted as a fact in macroecology, although it's still questioned in population dynamics. The size independence of S may also seem like a reasonable first assumption for ecosystems or the biota, but in fact it has worried many ecologists, because there is no known ecological mechanism that supports this assumption (e.g.,

Lawton, 1989). For example, what would happen if, on average, larger species had access to a wider range of resources or, by virtue of adaptations related to their size, were better at competing with small species? Then their equilibrium population sizes would be, on average, larger than the Damuth allometry would predict, and the exponents observed for the allometry of population density would be shallower than $-3/4$. What concerns ecologists is why it seems uncommon for communities or biotas to be assembled from species that together strongly violate the size independence of S. It may be even more surprising that species of one size or another do not evolve so as to violate the rough size independence of S for their trophic group. These questions remain unresolved, and the empirical scaling relationship—by now fairly well established—continues to be the subject of active interest (Enquist et al., 1998; Charnov et al., 2001; Blackburn and Gaston, 2001; Carbone and Gittleman, 2002).

Chapter 4

The maternal effect and/or predator–prey interactions remain contenders for explanation of the strong oscillatory effects we observe (i.e., those a few generations in length). The maternal-effect theory, developed in Ginzburg and Taneyhill (1994), Inchausti and Ginzburg (1998), and Ginzburg (1998), suggests periods of six or more generations. The period increases with declining maximum population growth rate. Observations of the periods (Krukouis and Schaffer, 1991) support this prediction. Predator–prey theory is also consistent with the observed data, but more parameters are used and the predictions are less specific compared with the maternal-effect model. A recent review of this debate by Turchin and Hanski (2001) argues in favor of the predation hypothesis, at least in the case of voles in Fennoscandia. They argue that the vole's annual maximal reproduction rate should be of the order of 100, whereas the maternal-effect model requires this value to be of the order of 30. This dispute might be resolved in the near future by empirical work.

In an old and not widely noticed paper, Michael Bulmer (1975) had suggested a simple but powerful test for determining the cause of population cycles. Bulmer pointed out that, based on a simple model, the phase shift (the delay) between the cycles of predator and prey can help us discern whether predator abundance is simply driven by food availability or if both species are engaged in joint dynamics. His conclusion was that the delay has to be greater or equal to 1/4 of the period of the cycle in order to support the claim that the cycle is driven by the interaction. If the delay is less than 1/4 of the period, it is more likely that the predator abundance is simply driven by prey abundance (following food availability). Putting aside issues concerning the precision in the value of 1/4 (which was deduced by linear approximation), the message is clear: it takes time for the predator–prey interaction to drive the prey abundance down from the maximum; it does not happen "instantaneously." On the other hand, if the predator abundance is driven by food, a generation (of the predator) after the food is at a maximum, we will expect the predator abundance to reach its maximum.

The Bulmer test is typically ignored by predator–prey modelers, one reason being that simultaneous data for both predator and prey are not usually available. Inchausti and Ginzburg (2002) analyzed the lag data for the lynx and hare. Their conclusion was that the lag was, on average, 1.5 years, clearly below 1/4 of the 10-year cycle period. This argues against predator–prey interactions as the cause of the cycles.

Vadasz et al. (2001, 2002) recently reported regular oscillations of the yeast cultures in the laboratory. This is definitely a single-species oscillation. The periods are reported to be about six to seven generations (unpublished personal communication). There is certainly more work required to demonstrate long cycles of single species in the laboratory. Such a demonstration would lend strong support for our point of view.

Chapter 5

The ratio-dependent predation model was developed independently by Ginzburg et al. (1971), Ginzburg et al. (1974), and Arditi (1975). The best known and most accessible account of this model is Arditi and Ginzburg (1989), which was followed by a large number of publications. The subject is covered now by major undergraduate textbooks. The review by Abrams and Ginzburg (2000) covers the debate and the current state of play. Two relevant papers published after the review are Skalski and Gilliam (2001) and Vucetich et al. (2002).

By analogy with "instantism" referring to the time scale, one can think of "localism." This would refer to ignoring spatial complexity and thinking of predators and prey as uniformly distributed. Various spatial structures give rise to different functional forms of predator dependence, commonly resembling ratio dependence (Cosner et al., 1999; Arditi et al., 2001).

The inequality (5.1) was stated without proof in Ginzburg et al. (1974) as a boundary between coexistence and dual extinction in ratio-dependent models. A number of recent papers contain comprehensive and detailed analyses of the the ratio-dependent predation model (see Jost et al., 1999; Berezovskaya et al., 2001; Hsu et al., 2001; Xiao and Ruan, 2001). These papers, particularly Berezovskaya et al. (2001), provide proofs of the inequality (5.1) even though it takes a bit of work to translate it into the form presented in chapter 5.

Chapter 6

Representing a hidden variable by the growth rate that, in addition to abundance, creates a two-dimensional description is an obvious idea—independent of the nature of the second, hidden dimension (Ginzburg and Inchausti, 1997). One advantage of this approach is that it makes data available for analysis based on just one data series. Another advantage is that the very existence of species-specific maximal growth rate produces two observable features: asymmetry of cycles and their limit-cycle character.

Linearization in terms of logarithmic acceleration, suggested in this chapter, can be restated in another commonly used language, that of the *superposition principle*. The issue of how one superimposes the effect of multiple food sources on a single consumer has recently attracted the attention of ecologists (Kareiva, 1994; Billick and Case, 1994; Wooton, 1994; Adler and Morris, 1994). According to the traditional view, the growth rate is a function of all the sources of food. It is linear in the Lotka–Volterra scheme and more complex in its various generalizations. In the simplest case, then, the suggestion is that the growth rate of a population with two food sources is the sum of two parts: the part due to consumption of the first resource plus that due to consumption of the second. This superposition principle is replaced in our model by the approximate additivity of acceleration contributions.

Superposition principles are equivalent to an assumption of linearity in a chosen space, and they have a long history in physics: from superimposing velocities in Aristotelian physics to Newtonian superposition of accelerations scaled by masses, to a bizarre superposition of the complex-valued functions in quantum physics and beyond. Although we cannot guess which form the superposition principle will take in the future of ecology, our suggestion at least sharpens the question in terms that may be worth considering.

Chapter 7

Ecological risk analysis has been a rapidly developing field in the last 20 years. Useful reference for further reading in this area are

Burgman et al. (1993), Akçakaya et al. (1999), and Ferson and Burgman (2000). Most methods in this area are implemented as software. The review by Pastorok et al. (2002) covers existing methods and software. The inability of the first-order models to capture the features that one might expect from a good theory for population management was a subject of an active discussion in Ginzburg (1992) and the comments and responses that followed.

Applications of theoretical ecology, traditionally, have been limited to single-species approaches, which pay a great deal of attention to the internal structure of populations and their spatial distribution. There are two reasons for this: (1) the relatively short time scale of the desired assessment and (2) our ability to better understand single-species dynamics. Most of the practical applications are implemented in software.

Chapter 8

Plato's famous cave analogy is presented in *The Republic* (Harvard Univ. Press, 1930), book 7. See Chalmers (1999) for a very nice discussion of the limitations of purely empirical elements in theory and hypothesis choice. Colyvan (2001a, chap. 4; 2002) argues that mathematics directly contributes to the aesthetic virtues of the theory; mathematics is not just a convenient language. This is relevant here because the choice of the mathematical model need not reduce to simply a matter of which model best corresponds to the underlying ecological facts. One model may be preferred for its aesthetic superiority, and this superiority might arise from the mathematics itself. Although this view of the role of mathematics in science is somewhat controversial, it is fair to say that the importance of mathematics in theory choice is underappreciated. This underappreciation seems to arise from (a perhaps understandable) overattention on the purely empirical elements of theory choice, and the widely held misconception that mathematics is *merely* the language of science. See Wigner

(1960), Dyson (1964), Steiner (1998), and Colyvan (2001b) for more on the role of mathematics in the progress of science.

Moving from the first- to the second-order description, or incorporating inertia, is certainly the simplest possible generalization one can make. Why, then, not the third or fourth order? Why not grandparental effect, in addition to the maternal effect? There certainly is evidence for it in a number of species, such as humans and our close evolutionary relatives. One argument is that the second-order description captures most, if not all, the features we need. In the linear approximation, a cubic characteristic equation has to have one real root. Therefore the dynamics will contain either a growing exponent (but we do not see unstable equilibria) or a declining exponent (these die out in time, and we do not see them for a different reason). If we saw the real presence of two different periods in our population abundance data, the fourth-order theory would have to be invoked to generate two distinct frequencies. Our data do testify to nonlinearity of population growth dynamics, but not to a more complex spectrum than a single frequency—at least on a multiple-generation time scale. Thus, the second-order is a logical approximation covering more, but far from all, of the dynamic properties we may need to include. The danger of overfitting is looming once again, and this is where aesthetic criteria may come into consideration.

Even though the satisfaction of certain aesthetic criteria is not a guarantee of being correct, satisfying such criteria at least serves as a temporary guide until the issue is settled by clear manipulative experiments. Note that the generality of an explanation is included by us in the list of aesthetic criteria, and this, too, might be challenged. Many people are much more comfortable thinking that every case has a different explanation; even the vole cycle for these authors is fundamentally different from that of the lemming cycle. This view reminds us again of Ptolemaic epicycles fitted separately and differently for every planet's trajectory.

Some of the overfitted models we have in mind are King and Shaffer (2001), Turchin and Batzli (2001), and Hanski et al. (2001).

There is another, stronger level of overfitting present in the Hanski et al. (2001) model. The functional response and the numerical response that is the coupling between predator and prey are assumed not to be linked (Ginzburg, 1998b). By making these two quantities functions of different arguments, one of N, the other of N/P, the authors acquire additional flexibility that may help in fitting (directly or indirectly), but this does not strike us as reasonable.

Stability selection (or, simply, *exclusion*) has certainly been a major force in shaping the solar system. Many bodies have fallen into the sun or collided with each other in the past. The resulting weakly interacting, and nested structure of planets and their satellites is a result of a long evolution under the influence of stability selection. Of course, biological structures are much younger than the solar system, but we may see the elements of ecological exclusion in the formation of trophic webs. Neutel et al. (2002) points toward certain simplifying tendencies in current trophic webs that are likely to be explained by ecological exclusion. Our suggestion of the inertial view of population growth and a simple planetary metaphor lies along similar lines. If proven useful by future work, it may contribute to substantial simplification of theories of how trophic webs work.

Apppendix B: Essential Features of the Maternal Effect Model

Even though the idea that the maternal effect could cause inertial population dynamics has existed since the 1950s, the new wave of interest in it appeared in the 1990s with the formal model suggested in Ginzburg and Taneyhill (1994). This model was at first applied to Lepidoptera because this was a simple case in which generations coincide with years and the data are available precisely in the generation time scale. Application to vole cycles followed (Inchausti and Ginzburg, 1998). It has been criticized in Turchin and Hanski (2001), which defended the more traditional predator–prey hypothesis as a cause of cyclicity. Here we briefly review the argument for the case of forest insects following the first publication as well as Ginzburg and Taneyhill (1995).

The equations describing the model are

$$N_{t+1} = N_t f(x_t)$$
$$x_{t+1} = g\left(x_t, \frac{S}{N_{t+1}}\right), \qquad \text{(B.1)}$$

where f is a monotonically increasing function of x_t and describes the net reproductive rate of an individual of quality x, and g is an increasing function of x (the maternal effect) and a decreasing function of N_{t+1} (as intraspecific competition for resources increases). Note that the argument N of the second equation is evaluated at the same generation as x on the left side of the equation. This is because quality is affected by density in the current generation, when competition for the resource takes place.

Mathematically, as we will show, this is a crucial assumption of the model. In this respect equations (B.1) differ from usual discrete-time population models where the variables N and x would be interpreted as densities of interacting populations (Beddington et al., 1976) and are evaluated at time t. Note that the model can be rewritten in the time-delay form by substituting the right side of the first equation for N_{t+1} in the second equation and then eliminating x_t. It will then look like $N_{t+1} = F(N_t, N_{t+1})$.

In the absence of the maternal effect g is independent of x, and equations (B.1) reduce to the standard form of the immediate density-dependent model: $N_{t+1} = N_t F(N_t)$. The presence of the maternal effect thus creates the delayed density dependence. Equilibrium density and quality (N^*, x^*) can be found as the roots of

$$f(x^*),$$

$$g(x^*, N^*) = x^*.$$

The model given by equations (B.1) can then be analyzed by the method of local stability analysis. Let us first change the variables to logarithmic scale:

$$u = \ln N,$$

$$v = \ln x,$$

and define new functions

$$f(v) = \ln f(e^v),$$

$$g(v, u) = \ln(e^v, e^u).$$

We thus have in the new variables

$$u_{t+1} = u_t + f(v_t),$$

$$v_{t+1} = g(v_t, u_{t+1}). \tag{B.2}$$

The matrix for the linearized equations around the equilibrium point is

$$\begin{bmatrix} 1 & f'_v \\ g'_u & g'_v + g'_u f'_v \end{bmatrix},$$

where prime denotes the partial derivative. Note that the argument of the second equation in (B.2) is u_{t+1}, and it should be substituted from the first equation when computing derivatives. Stability of the system is determined by the roots of the characteristic equation

$$\lambda^2 - \left(1 + g'_v + g'_u f'_v\right)\lambda + g'_v = 0.$$

Local stability is governed by, in the original variables,

$$a = g'_x > 0$$

and

$$b = -f'_x g'_N N^* > 0.$$

In order to simplify the analysis of (B.1), it will be convenient to discuss a specific example. Consider the following parameterization of functions f and g:

$$N_{t+1} = N_t R \frac{x_t}{k + x_t}$$

$$x_{t+1} = x_t M \frac{S/N_{t+1}}{p + S/N_{t+1}} \quad \text{(B.3)}$$

The parameter R represents the maximum reproductive rate given any quality x, and M is the maximum possible increase in average quality. Constants k and p control the rates of increase to the asymptotes R and M. The S term represents the total amount of resource available in the environment; we assume that this is constant each generation. We divide the numerator and denominator of equations (B.3) by S; then, by expressing N and x in the appropriate units, we can eliminate the other two parameters k and p so that equations (B.3) in the dimensionless form become

$$N_{t+1} = N_t R \frac{x_t}{1 + x_t},$$

$$x_{t+1} = x_t M \frac{M}{1 + N_{t+1}} \qquad (B.4)$$

Stability analysis of equations (B.4) shows that nondamping oscillations occur whenever the parameter R is greater than unity, assuming that M is also greater than unity. The period of oscillations in the linearized form of equations (B.4) is a function of

$$b = (1 - 1/R)(1 - 1/M).$$

Thus, if $M \gg 1$, the length of the period is determined only by the maximum rate of increase R. In this model, low values of R lead to longer cycles.

Figure B1 shows a typical behavior of model (B.4) displayed as a phase-plane plot. As can be seen from the phase diagram, the cycles are neutrally stable; that is, the amplitude and period of the cycles depend upon initial conditions. In this way, equations (B.4) are similar to the familiar Lotka–Volterra predator and prey equations, and also to several other two-dimensional models (Beddington et al., 1975; Anderson and May, 1980; Lauwerier and Metz, 1986). The cycles usually have fractional values for the period, and the system is not precisely periodic, as in continuous models. Note also in figure B.1 the formation of island chains for some initial conditions, a feature common to this type of two-dimensional discrete map (Lauwerier, 1986; Lauwerier and Metz, 1986).

Models that generate neutral cycles are of course pathological from a biological viewpoint (May, 1974b). Modification of

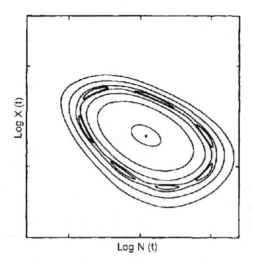

Figure B.1. Typical phase-plane portrait of the maternal-effect model, shown in logarithmic scale. Each ellipse corresponds to a unique set of initial conditions. Note the formation of "island chains" for some initial values (Ginzburg and Taneyhill, 1994). Reprinted with permission of the *Journal of Animal Ecology*.

equations (B.4) to produce true *limit cycles* can be done in any number of ways by adding one or more parameters to the model. Available data do not allow us to distinguish between different modifications, but our main conclusions do not depend on the exact form of the functions.

Model (B.4) and the Lauwerier model have several properties in common. First, and perhaps most important, is that cycle periods have an absolute minimum of six. This result was stated as a theorem by Lauwerier and Metz (1986) for the simpler host–parasitoid model, which mathematically is a special case of our model (B.1). Our formal result is slightly more general. The minimum six-generation cycle is in perfect agreement with observation: to our knowledge, no forest moth population has a period less than 6 (Myers, 1988).

Second, cycle periods decrease as the growth parameters (R) increase. This pattern is opposite to that of some other delay models, such as the discrete predator-prey (Beddington et al., 1975) or delayed logistic (Levin and May, 1976) models (see the comparison in Taneyhill, 1993). The limit of six generations corresponds to high growth rates. On the other extreme, when R is close to 1, very long periods are possible. The single main qualitative evidence in favor of our model comes if we assume $M \gg 1$ and thus present the period of cycles only as a function of R (figure B.2).

The data evaluated for different annual insects by us and by Berryman (1995) appear to fit the predicted curve quite well. R values per generation are not very different for small mammals and insects, so the range of periods for mammalian cycles in generation time units serves as a confirmation of the proposed view.

Note that the theoretical curve and the limit of six generations are only valid for local dynamics around the equilibrium. Limit

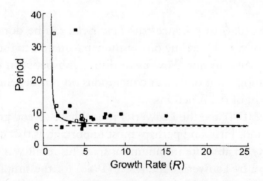

Figure B.2. Comparison of Berryman (1995) data (solid squares) and Ginzburg and Taneyhill (1994) data (open squares). The curve is from Ginzburg and Taneyhill (1995). Horizontal line corresponds to a theoretical lower limit of six generations. Reprinted with permission of the *Journal of Animal Ecology*.

cycles and chaotic trajectories surrounding unstable equilibria have periods or quasi periods exceeding the linearized period, and this is one reason why most real data points lie above and to the right of the line.

As we pointed out in chapter 4, the essence of the argument lies in the combination of delayed density dependence, the assumption of the per capita consumption of the resource, S/N, and the correct choice of timing. This corresponds in logarithmic scale to the second difference relating negatively to abundance, as in Hooke's law for the oscillating mass attached to the spring. The coefficient, or Hooke's constant, does not exceed unity, and this causes the period to be longer than six generations. Various generalizations of the model discussed in chapter 4 only elongate the period.

We thus suggest that every species possesses an eigenperiod of longer than six of its generations. Whether or not a given population cycles, and with what amplitude and shape, may depend on disturbances. Predator–prey interaction is one of the prime candidates to cause such a disturbance. However, if a population cycles, the period, in our view, is an eigenperiod.

Bibliography

Abrams, P. A. 1994. The fallacies of ratio-dependent predation. *Ecology* 74: 1842–1850.

Abrams, P. A., and Ginzburg, L. R. 2000. The nature of predation: prey dependent, ratio dependent, or neither? *Trends in Ecology and Evolution* 15(8): 337–341.

Abrams, P. A., Menge, B. A., Mittelbach, G. G., Spiller, D. A., and Yodzis, P. 1996. The role of indirect effects in food webs. In G. A. Polis and K. O. Winemiller (eds.), *Food Webs: Integration of Patterns and Dynamics*. New York: Chapman and Hall, p. 387.

Adler, F. R., and Morris, W. F. 1994. A general test for interaction modification. *Nature*. 75(6): 1552–1559.

Akçakaya, H. R. 1992. Population cycles of mammals: evidence for a ratio-dependent predation hypothesis. *Ecological Monographs* 62: 119–142.

Akçakaya, H. R., Arditi, R., and Ginzburg, L. R. 1995. Ratio-dependent predation: an abstraction that works. *Ecology* 76(3): 995–1004.

Akçakaya, H. R., Burgman, M. A., and Ginzburg, L. R. 1999. *Applied Population Ecology: Principles and Computer Exercises*. Sunderland, MA: Sinauer Associates.

Akçakaya, H. R., Ginzburg, L. R., Slice, D., and Slobodkin, L. B. 1988. The theory of population dynamics—II. Physiological delays. *Bulletin of Mathematical Biology* 50(5): 503–515.

Anderson, R. M., and May, R. M. 1980. Infectious diseases and population cycles of forest insects. *Science* 210: 658–661.

Arditi, R. 1975. Etude de modeles theoriques de l'ecosysteme a deux especes proie predateur. Specialty Doctorate Thesis, University of Paris.

Arditi, R., and Ginzburg, L. R. 1989. Coupling in predator–prey dynamics: ratio-dependence. *Journal of Theoretical Biology* 139: 311–326.

Arditi, R., Tyutyunov, Y., Morgulis, A., Govorukhin, V., and Senina, I. 2001. Directed movement of predators and the amergence of density-dependence in predator–prey models. *Theoretical Population Biology* 59: 207–221.

Bibliography

Armstrong, D. M. 1983. *What Is a Law of Nature?* Cambridge: Cambridge University Press.

Baltensweiler, W., and Fischlin, A. 1989. The larch budmoth in the Alps. In A. Berryman (ed.), *Dynamics of Forest Insect Populations: Patterns, Causes and Implications*. New York: Plenum Press, pp. 331–351.

Beckerman, A., Benton, T. G., Ranta, E., Kaitala, V., and Lundberg, P. 2002. Population dynamic consequences of delayed life-history effects. *Trends in Ecology and Evolution* 17: 263–269.

Beddington, J. R., Free, C. A., and Lawton, J. H. 1975. Dynamic complexity in predator–prey models framed in difference equations. *Nature* 225: 58–60.

Beddington, J. R., Free, C. A., and Lawton, J. H. 1976. Concepts of stability and resilience in predator–prey models. *Journal of Animal Ecology* 45: 791–816.

Bell, E. T. 1953. *Men of Mathematics*. London: Penguin.

Berezovskaya, F., Karev, G., and Arditi, R. 2001. Parametric analysis of the ratio-dependent predator-prey model. *Journal of Mathematical Biology* 43: 221–246.

Berryman, A. A. 1995. Population cycles: a critique of the maternal and allometric hypotheses. *Journal of Animal Ecology* 64: 290–293.

Berryman, A. A. 1999. *Principles of Population Dynamics and Their Application*. Cheltenham, UK: Stanley Thornes Publishers.

Berryman, A. A. (ed.) 2002. *Population Cycles: The Case for Trophic Interactions*. New York: Oxford University Press.

Billick, I., and Case, T. J. 1994. Higher order interactions in ecological communities: what are they and how can they be detected? *Nature* 75(6): 1529–1543.

Bjornstad, O. N., and Grenfell, B. 2001. Noisy clockwork: time series analysis of population fluctuations in animals. *Science* 293:638–643.

Blackburn, T. M., and Gaston, K. J. 2001. Linking patterns in macroecology. *Journal of Animal Ecology* 70: 338–352.

Bonner, J. T. 1965. *Size and Cycle*. Princeton, NJ: Princeton University Press.

Boonstra, R. 1994. Population cycles in microtines: the senescene hypothesis. *Evolutionary Ecology* 8: 196–219.

Boonstra, R., and Boag, P. T. 1987. A test of the Chitty hypothesis: inheritance of life-history traits in meadow voles Microtus pennsylvanicus. *Evolution* 41: 929–947.

Boonstra, R., Krebs, C. J., and Stenseth, N. C. 1998. Population cycles in small mammals: the problem of explaining the low phase. *Ecology* 79(5): 1479–1488.

Brown, J. H. 1995. *Macroecology*. Chicago: University of Chicago Press.

Brown, J. H., and West, G. (eds.) 2000. *Scaling in Biology*. New York: Oxford University Press.

Brown, J. H., West, G. B., and Enquist, B. J. 2000. Scaling in biology: patterns and processes, causes and consequences. In J. H. Brown and G. B. West (eds.), *Scaling in Biology*. Oxford: Oxford University Press, pp. 1–24.

Bulmer, M. 1975. Phase relations in the ten-year cycle. *Journal of Animal Ecology* 44: 609–621.

Burgman, M. A., Ferson, S., and Akçakaya, H. R. 1993. *Risk Assessment in Conservation Biology*. London: Chapman and Hall.

Calder, W. A. 1983. An allometric approach to population cycles of mammals. *Journal of Theoretical Biology* 100: 275–282.

Calder, W. A., III. 1984. *Size, Function, and Life History*. Mineola, NY: Dover edition, 1996.

Carbone, C., and Gittleman, J. L. 2002. A common rule for the scaling of carnivore density. *Science* 295: 2273–2276.

Cartwright, N. 1983. *How the Laws of Physics Lie*. New York: Oxford University Press.

Cartwright, N. 1999. *The Dappled World: A Study of the Boundaries of Science*. Cambridge: Cambridge University Press.

Chalmers, A. F. 1999. *What Is This Thing Called Science?* 3d ed. Brisbane: University of Queensland Press.

Charnov, E. L. 1993. *Life History Invariants: Some Explorations of Symmetry in Evolutionary Ecology*. Oxford: Oxford University Press.

Charnov, E. L., Haskell, J., and Ernest, S. K. M. 2001. Density-dependent invariance, dimensionles life histories and the energy-equivalence rule. *Evolutionary Ecology Research* 3: 117–127.

Clark, G. P. 1971. The second derivative in population modelling. *Ecology* 52: 606–613.

Clutton-Brock, T. H., Albon, S., and Guinness, F. 1987. Interactions between population density and maternal characteristics affecting fecundity and juvenile survival in red deer. *Journal of Animal Ecology* 56: 857–871.

Colyvan, M. 2001a. *The Indispensability of Mathematics*. New York: Oxford University Press.

Colyvan, M. 2001b. The miracle of applied mathematics. *Synthese* 127(3): 265–278.

Colyvan, M. 2002. Mathematics and aesthetic considerations in science. *Mind* 111(441): 69–74.

Colyvan, M., and Ginzburg, L. R. 2003a. The Galilean turn in population ecology. *Biology and Philosophy*, 18(3): 401–414.

Colyvan, M., and Ginzburg, L. R. 2003b. Laws of nature and laws of ecology. *Oikos* 101(3): 649–653.

Cooper, G. 1998. Generalizations in ecology: a philosophical taxonomy. *Biology and Philosophy* 13: 555–586.

Cosner, C., DeAngelis, D. L., Ault, J. S., and Olson, D. B. 1999. Effects of spatial grouping on the functional response of predators. *Theoretical Population Biology* 56: 65–75.

Damuth, J. 1981. Population density and body size in mammals. *Nature* 290: 699–700.

Damuth, J. 1987. Interspecific allometry of population density in mammals and other animals: the independence of body mass and population energy-use. *Biological Journal of the Linnean Society* 31: 193–246.

Damuth, J. 1991. Ecology: of size and abundance. *Nature* 351: 268–269.

Duhem, P. 1906. *The Aim and Structure of Physical Theory.* Princeton, NJ: Princeton University Press, 1954.

Dunbar, M. J. 1971. The evolutionary stability in marine environments: natural selection at the level of the ecosystem. In G. C. Williams (ed.), *Group Selection.* Chicago: Aldine Atherton, 125–135.

Dyson, F. J. 1964. Mathematics in the physical sciences. *Scientific American* 211(3): 128–146.

Easterlin, R. 1961. The American baby-boom in historical perspective. *American Economic Review* 51: 860–911.

Elton, C. 1927. *Animal Ecology.* Chicago: University of Chicago Press, 2001.

Elton, C., and Nicholson, M. 1942. The ten year cycle in numbers of lynx in Canada. *Animal Ecology* 2: 215–244.

Enquist, B. J., Brown, J. H., and West, G. B. 1998. Allometric scaling of plant energetics and population density. *Nature* 395: 163–165.

Fenchel, T. 1974. Intrinsic rate of natural increase: the relationship with body size. *Oecologia* 14: 317–326.

Ferson, S., and Burgman, M. A. (eds.) 2000. *Quantitative Methods for Conservation Biology.* New York: Springer-Verlag.

Feynman, R. 1965. *The Character of Physical Law.* London: British Broadcasting Corporation.

Feynman, R., Leighton, R., and Sands, M. 1963. *The Feynman Lectures on Physics.* Reading, MA: Addison-Wesley.

Frauenthal, J. 1975. A dynamic model for human population growth. *Theoretical Population Biology* 8: 64–73.

Ginzburg, L. R. 1972. The analogies of the "free motion" and "force" concepts in population theory (in Russian). In V. A. Ratner (ed.), *Studies on Theoretical Genetics.* Novosibirsk, USSR: Academy of Sciences of the USSR, pp. 65–85.

Ginzburg, L. R. 1986. The theory of population dynamics: I. Back to first principles. *Journal of Theoretical Biology* 122: 385–399.

Ginzburg, L. R. 1992. Evolutionary consequences of basic growth equations. *Trends in Ecology and Evolution* 7: 133; correspondence, 314–317; further letters, 1993, 8: 68–71.

Ginzburg, L. R. 1998. Inertial growth: population dynamics based on maternal effects. In T. A. Mousseau and C. W. Fox (eds.), *Maternal Effects as Adaptions*. New York: Oxford University Press, pp. 42–53.

Ginzburg, L. R., and Akçakaya, H. R. 1992. Consequences of ratio-dependent predation for steady state properties of ecosystems. *Ecology* 73(5): 1536–1543.

Ginzburg, L. R., Akçakaya, H. R., Slice, D., and Slobodkin, L. B. 1988. Balanced growth rates vs. balanced accelerations as causes of ecological equilibria. In L. M. Ricciardi (ed.), *Biomathematics and Related Computational Problems*. The Netherlands: Kluwer Academic Publishers, pp. 165–175.

Ginzburg, L. R., Goldman, Y. I., and Rahkin, A. I. 1971. A mathematical model of interaction between two populations I. Predator prey (in Russian). *Zhurnal Obshchei Biologii* 32: 724–730.

Ginzburg, L. R., and Inchausti, P. 1997. Asymmetry of population cycles: abundance-growth representation of hidden causes of ecological dynamics. *Oikos* 80: 435–447.

Ginzburg, L. R., and Jensen, C. X. J. (forthcoming). Rules of thumb for judging ecological theories. *Trends in Ecology and Evolution*.

Ginzburg, L. R., Konovalov, N. Y., and Epelman, G. S. 1974. A mathematical model of interaction between two populations. IV. Comparison of theory and experimental data (in Russian). *Zhurnal Obshchei Biolgii* 35: 613–619.

Ginzburg, L. R., and Taneyhill, D. E. 1994. Population cycles of forest Lepidoptera: a maternal effect hypothesis. *Journal of Animal Ecology* 63: 79–92.

Ginzburg, L. R., and Taneyhill, D. 1995. Higher growth rate implies shorter cycle, whatever the cause: a reply to Berryman. *Journal of Animal Ecology* 64: 294–295.

Hanski, I., Henttonen, H., Korpimaki, E., Oksanen, L., and Turchin, P. 2001. Small-rodent dynamics and predation. *Ecology* 82(6): 1505–1520.

Harvey, P. H., and Pagel, M. D. 1991. *The Comparative Method of Evolutionary Biology*. New York: Oxford University Press.

Hemmingsen, A. M. 1960. Energy metabolism as related to body size and respiratory surfaces, and its evolution. Reports of the Steno Memorial Hospital and the Nordisk Insulin Labatorium. Volume IX, Part II. Princeton, NJ: Novo Nordisk, pp. 6–110.

Hempel, C. G. 1965. *Aspects of Scientific Explanation and Other Essays in the Philosophy of Science*. London: Macmillan.

Henttonen, H., Oksanen, L., Jortikka, T., and Hauksalami, V. 1987. How much do weasels shape microtine cycles in the northern Fennoscandian taiga? *Oikos* 50: 353–365.

Holland, M. M., Lawrence, D. M., and Morrin, D. J. (eds.) 1992. *Profiles of Ecologists: Results of a Survey of the Membership of the Ecological Society of America.* Washington, DC: Ecological Society of America.

Hörnfeldt, B. 1994. Delayed density dependence as a determinant of vole cycles. *Ecology* 75: 791–806.

Hsu, S.-B., Hwang, T.-W., and Kuamg, Y. 2001. Global analysis of the Michaelis-Menten-type ratio-dependent predator–prey system. *Journal of Mathematical Biology* 42: 489–506.

Inchausti, P., and Ginzburg, L. R. 1998. Small mammal cycles in northern Europe: pattern and evidence for a maternal effect hypothesis. *Journal of Animal Ecology* 67: 180–194.

Inchausti, P., and Ginzburg, L. R. 2002. Using the phase shift for assessing the causation of population cycles. *Ecological Modeling* 152: 89–102.

Inchausti, P., and Halley, J. 2001. Investigating long-term ecological variability using the global population dynamics database. *Science* 293: 655–657.

Jost, C., Arino, O., and Arditi, R. 1999. About deterministic extinction in ratio-dependent predator–prey models. *Bulletin of Mathematical Biology* 61: 19–32.

Jost, C., and Ellner, S. P. 2000. Testing for predator dependence in predator–prey dynamics: a non-parametric approach. *Proceedings of the Royal Society of London. Series B. Biological Sciences* 267(1453): 1611–1620.

Kareiva, P. M. 1994. Higher order interactions as a foil to reductionist ecology. *Nature* 75(6): 1527–1528.

King, A. A., and Shaffer, W. 2001. The geometry of a population cycle: a mechanistic model of snowshoe hare demography. *Ecology* 82(3): 814–830.

Kingsland, S. E. 1985. *Modelling Nature: Episodes in the History of Population Ecology.* Chicago: University of Chicago Press.

Kitcher, P. 1981. Explanatory unification. *Philosophy of Science* 48: 507–531.

Kleiber, M. 1975. *The Fire of Life*, 2d ed. New York: Krieger.

Krebs, C. J., Boonstra, R., Boutin, S., and Sinclair, A. R. E. 2001a. What drives the 10-year cycle of snowshoe hares? *Bioscience* 51(1): 25–35.

Krebs, C. J., Boutin, S., and Boonstra, R. (eds.) 2001b. *Ecosystem Dynamics of the Boreal Forest: The Kluane Project.* New York: Oxford University Press.

Krebs, C. J., Boutin, S., Boonstra, R., Sinclair, A. R. E., Smith, J. N. M., Dale, M. R. T., Martin, K., and Turkington, R. 1995. Impact of food and predation on the snowshoe hare cycle. *Science* 269: 1112–1115.

Krukonis, G., and Schaffer, W. M. 1991. Population cycles in mammals and birds: does periodicity scale with body size? *Journal of Theoretical Biology* 148: 469–493.

Lakatos, I. 1970. Falsification and the methodology of scientific research programmes. In I. Lakatos and A. Musgrave (eds.), *Criticism and the Growth of Knowledge*. Cambridge: Cambridge University Press, pp. 91–195.

Lauwerier, H. A. 1986. Two-dimensional iterative maps. In A. V. Holden (ed.), *Chaos*. Princeton, NJ: Princeton University Press, chap. 4.

Lauwerier, H. A., and Metz, J. A. J. 1986. Hopf bifurcation in host-parasitoid models. *Journal of Mathematics in Applied Medicine and Biology* 3: 191–210.

Lawton, J. H. 1989. What is the relationship between population density and body size in animals? *Oikos* 55: 429–434.

Lawton, J. H. 1999. Are there general laws in biology? *Oikos* 84: 177–192.

Luckinbill, L. S. 1973. Coexistence in laboratory populations of paramecium aurelia and its predator didinium nasutum. *Ecology* 54(6): 1320–1327.

MacArthur, R. 1972. Coexistence of species. In J. A. Behnke (ed.), *Challenging Biological Problems: Directions towards Their Solution*. New York: Oxford University Press, pp. 253–259.

May, R. M. 1974a. *Stability and Complexity in Model Ecosystems*, 2d ed. Princeton, NJ: Princeton University Press.

May, R. M. 1974b. Biological populations with nonoverlapping generations: stable cycles and chaos. *Science* 186: 645–647.

McCarthy, M. A., Ginzburg, L. R., and Akçakaya, H. R. 1995. Predator interference across trophic chains. *Ecology* 76(4): 1310–1319.

McCauley, E., and Murdoch, W. 1985. Cyclic and stable populations: plankton as a paradigm. *American Naturalist* 129: 97–121.

Middleton, A. 1934. Periodic fluctuations in British game populations. *Journal of Animal Ecology* 3: 231–249.

Millar, J. S., and Zammuto, R. M. 1983. Life histories of mammals: an analysis of life tables. *Ecology* 64(4): 631–635.

Morin, P. J. 1999. *Community Ecology*. Malden: Blackwell Science.

Mousseau, T. A., and Fox, C. W. (eds.) 1998. *Maternal Effects as Adaptions*. New York: Oxford University Press.

Murdoch, W. W. 1970. Population regulation and population inertia. *Ecology* 51(3): 497–502.

Murdoch, W. W., Briggs, C., and Nisbet, R. 2003. *Consumer-Resource Dynamics*. Princeton, NJ: Princeton University Press.

Murdoch, W. W., Kendall, B. E., Nisbet, R. M., Briggs, C. J., McCauley, E., and Bolser, R. 2002. Single-species models for many-species food webs. *Nature* 417: 541–543.

Murray, B. M., Jr. 1992. Research methods in physics and biology. *Oikos* 64(3): 594–596.

Murray, B. M., Jr. 1999. Is theoretical ecology a science? *Oikos* 87: 594–600.

Myers, J. H. 1988. Can a general hypothesis explain population cycles of forest Lepidoptera? *Advances in Ecological Research* 18: 179–242.

National Environment Research Council (NERC). 1999. The Global Population Dynamics Database. NERC Centre for Population Biology, Imperial College. Available: http://www.sw.ic.ac.ul/cpb/cpb/gpdd.html. Accessed 6/13/2002.

Nee, S., Read, A. F., Greenwood, J. J. D., and Harvey, P. H. 1991. The relationship between abundance and body size in British birds. *Nature* 351: 312–313.

Neutel, A., Heesterbeek, J., and de Ruiter, P. 2002. Stability in real food webs: weak links in long loops. *Science* 296: 1120–1123.

Newton-Smith, W. H. 2000. *A Companion to the Philosophy of Science*. Oxford: Blackwell.

Pastorok, R. A., Bartell, S. M., Ferson, S., and Ginzburg, L. R. (eds.) 2002. *Ecological Modeling in Risk Assessment*. Boca Raton: CRC Press.

Peek, J., Urich, D., and Mackie, R. 1976. Moose habitat selection and relationship to forest management in northeastern Minnesota. *Wildlife Monographs* 48: 6–65.

Peterson, R. O., Page, R. E., and Dodge, K. M. 1984. Wolves, moose, and the allometry of population cycles. *Science* 224(4655): 1350–1352.

Peterson, R. O., and Vucetich, J. A. 2002. *Ecological studies of Wolves on Isle Royal, Annual Report 2001–2002*. Houghton, MI: School of Forestry and Wood Products, Michigan Technological University.

Plato (P. Shorey, trans.) 1930. *The Republic*. Cambridge, MA: Harvard University Press.

Purcell, E. M. 1977. Life at low Reynolds number. *American Journal of Physics* 45(1): 3–11.

Quenette, P. Y., and Gerard, J. F. 1993. Why biologists do not think like physicists. *Oikos* 68(2): 361–363.

Quine, W. V. 1951. Five milestones of empiricism. In *Theories and Things*. Cambridge, MA: Harvard University Press, 1981, pp. 67–72.

Quine, W. V. 1960. *Word and Object*. New York: MIT Press and John Wiley and Sons.

Quine, W. V. 1980. Two dogmas of empiricism. In *From a Logical Point of View*, 2d ed. Cambridge, MA: Harvard University Press, pp. 20–46.

Quine, W. V. 1995. *From Stimulus to Science*. Cambridge, MA: Harvard University Press.

Quine, W. V., and Ullian, J. S. 1978. *The Web of Belief*, 2d ed. New York: McGraw-Hill.

Reiss, M. H. 1989. *The Allometry of Growth and Reproduction.* Cambridge: Cambridge University Press.

Rossiter, M. C. 1991. Environmentally based maternal effects: a hidden force in population dynamics. *Oecologia* 87: 288–294.

Rossiter, M. 1998. The role of environmental variation in parental effects expression. In T. A. Mousseau and C. W. Fox (eds.), *Maternal Effects as Adaptations.* New York: Oxford University Press, pp. 113–134.

Royama, T. 1992. *Analytical Population Dynamics.* London: Chapman and Hall.

Schaffer, W. M., Pederson, B. S., Moore, B. K., Skarpaas, O., King, A. A., and Bronnikova, T. V. 2001. Sub-harmonic resonance and multi-annual oscillations in northern mammals: a non-linear dynamical systems perspective. *Chaos Solutions and Fractals* 12(2): 251–264.

Seldal, T., Amndersen, K., and Hodgstedt, G. 1994. Grazing-induced proteins inhibitors: a possible cause for lemming population cycles. *Oikos* 70: 3–11.

Skalski, G. T., and Gilliam, J. F. 2001. Functional responses with predator interference: viable alternatives to the Holling type II model. *Ecology* 82(11): 3083–3092.

Slobodkin, L. 2003. *A Citizen's Guide to Ecology.* New York: Oxford University Press.

Slobodkin, L. B. 1961. *Growth and Rejuvenation of Animal Populations*, 2d ed. New York: Dover, 1980.

Smart, J. J. C. 1968. *Between Philosophy and Science: An Introduction to the Philosophy of Science.* New York: Random House.

Sobel, D. 1999. *Galileo's Daughter: A Historical Memoir of Science, Faith, and Love.* New York: Walker and Company.

Southwood, T. R. E. 1976. Bionomic strategies and population parameters. In R. M. May (ed.), *Theoretical Ecology: Principles and Applications.* Philadelphia: Saunders, pp. 26–48.

Spencer, D., and Lensink, C. 1970. The muskox of the Nunivak Island, Alaska. *Journal of Wildlife Management* 34: 1–15.

Steiner, M. 1998. *The Applicability of Mathematics as a Philosophical Problem.* Cambridge, MA: Harvard University Press.

Stenseth, N. C. 1999. Population cycles in voles and lemmings: density dependence and phase dependence in a stochastic world. *Oikos* 87(2): 427–461

Taneyhill, D. 1993. The logistic equation: final installment. *Trends in Ecology and Evolution* 8: 68–70.

Turchin, P. 2001. Does population ecology have general laws? *Oikos* 94(1): 17–26.

Turchin, P. 2003a. *Complex Population Dynamics: A Theoretical/Empirical Synthesis.* Princeton, NJ: Princeton University Press.

Turchin, P. 2003b. *Historical Dynamics: Why States Rise and Fall.* Princeton, NJ: Princeton University Press.

Turchin, P., and Batzli, G. 2001. Availability of food and the population dynamics of arvicoline rodents. *Ecology* 82: 1521–1534.

Turchin, P., and Hanski, I. 2001. Contrasting alternative hypotheses about rodent cycles by translating them into parameterized models. *Ecology Letters* 4: 267–276.

Vadasz, A. S., Vadasz, P., Abashar, M. E., and Gupthar, A. S. 2001. Recovery of an oscillatory mode of batch yeast growth in water for a pure culture. *International Journal of Food Microbiology* 71(2–3): 219–234.

Vadasz, A. S., Vadasz, P., Abashar, M. E., and Gupthar, A. S. 2002. Theoretical and experimental recovery of oscillations during batch growth of a mixed culture of yeast in water. *World Journal of Microbiology and Biotechnology* 18(3): 239–246.

van Fraassen, B. C. 1980. *The Scientific Image.* Oxford: Clarendon.

van Fraassen, B. C. 1989. *Laws and Symmetry.* Oxford: Clarendon.

Veilleux, B. G. 1979. An analysis of the predatory interaction between paramecium and didinium. *Journal of Animal Ecology* 48(3): 787–803.

Vucetich, J. A., Peterson, R. O., and Schaefer, C. L. 2002. The effect of prey and predator densities on wolf predation. *Ecology* 83: 3003–3013.

Wellington, W. G. 1957. Individual differences as a factor in population dynamics: the development of a problem. *Canadian Journal of Zoology* 35: 293–323.

West, G. B., Brown, J. H., and Enquist, B. J. 1999. The fourth dimension of life: fractal geometry and allometric scaling of organisms. *Science* 284: 1677–1679.

Wigner, E. P. 1960. The unreasonable effectiveness of mathematics in the natural sciences. *Communications on Pure and Applied Mathematics* 13: 1–14.

Wooton, J. T. 1994. Putting the pieces together: testing the independence of interactions among organisms. *Nature* 75(6): 1544–1551.

Xiao, D., and Ruan, S. 2001. Global dynamics of a ratio-dependent predator-prey system. *Journal of Mathematical Biology* 43(3): 268–290.

Index

abundance, average, 113
abundance equilibrium, 43–46, 46–48, 107–109, 113
accelerated death, 35–36
   and analogy with falling bodies, 36–39, 41, 42–43
   in contrast to exponential death, 35–36
   experimental verification of, 39–42
   and metabolism, 43–46
aesthetics in science, 30, 100, 120–123
Akçakaya, H. R., 68
analogies, scientific, 8–9, 57
   falling bodies and accelerated death, 36–39, 41, 42–43
   falling body on a spring and population equilibrium, 46
   friction and limiting factors in population growth, 98–99
   Galilean relativity principle and ecological invariance, 96–97
   mechanical oscillations and cyclic population growth, 59–62, 77, 89
   Newtonian mechanics and inertial population growth, 100–103
   physics and ecology, 3–4, 8–9, 57

planetary motion/orbits and population growth, 3–10, 57
Aristotle, views on motion, 4–5, 37

body size
   and relationship with basal metabolic rate, 12–15, 20, 21, 31, 46
   and relationship with generation time, 16, 19–20
   and relationship with maximal reproduction rate, 17, 19–20, 21
   and relationship with population density, 17–18, 46–48, 102
   and relationship with population oscillations, 18–19, 21, 57–59, 74, 102
Bonner, J. T., 16
Bos taurus, 17
Brahé, Tycho, 21

Calder, William, 18, 20, 57
Calder allometry, 18–19, 21, 57–59, 74, 102
carrying capacity, 34. *See also* abundance equilibrium
Cervus elaphus, 92
Clethrionomys rufocanus, 93
coexistence, predator and prey, 67–69, 109, 112

cohort effect, 55
cyclic population growth, 6–8, 50–52, 52–57. See also inertial population growth
  and age structure, 55
  and analogy with mechanical oscillations, 59–62, 77, 89
  and cohort effect, 55
  cycle periods of, 52–57, 57–59, 59–62, 62–63, 79–82
  of *Daphnia* spp., 53
  Easterlin hypothesis of, 56
  Eigenperiod hypothesis of, 59–62
  external mechanisms of, 55, 79
  in fisheries, 107, 115
  of hares, 53, 55, 60–61
  of humans, 53, 56
  of insects, 53, 55
  internal mechanisms of, 55, 74, 79 (see also maternal effect)
  of lemmings, 55, 60, 61, 79, 93
  of *Lepidoptera* species, 53, 83
  and lynx-hare cycle, 55, 60–61, 68–69
  and maternal effect, 50–52, 52–57, 78–82
  of moose, 53
  and number of model parameters, 124–125
  and predator–prey interactions, 57–59, 64–82
  and second-order models, 54, 79, 84–89
  time scales of, 51, 71–74, 87, 106–107, 130
  of voles, 53, 60, 61, 79, 92, 93

Damuth, John, 15, 17, 46
Damuth allometry, 17–18, 46–48, 102
  for carnivores, 47, 48
  for frugivores, 47
  for herbivores, 47
  for insectivores, 47
Darwin, Charles, 8
delayed density dependence, 51, 94, 108
differential equations, 90
  and correspondence to models, 85–86
  and idealization of time, 71–74, 85–86
  parametric specification for, 85–90
  second-order, 44–46, 84–89
  Taylor-series expansions for, 90–91
donor-controlled models, 75–77
dual extinction of predator and prey, 67–69, 112

Easterlin hypothesis, 56
ecological allometries, 12, 19–21, 25
  Calder allometry, 18–19, 21, 57–59, 74, 102
  Damuth allometry, 17–18, 46–48, 102
  Fenchel allometry, 17, 19–20, 21
  generation-time allometry, 16, 19–20
  Kleiber allometry, 12–15, 20, 21, 31, 46
ecological exclusion, 112
ecological forces, 101–103
eigenfrequency, 60
eigenperiod hypothesis, 59–62
Einstein, Albert, 79, 119, 122
electromagnetic theory, 8
Eltonian pyramid, 127, 129
Escherichia coli, 35
evolution rates, 111–113
exponential death, contrasted with accelerated death, 35–36

exponential population growth, 7–8,
    35–36, 96–98, 129
  as first law of ecology, 101, 129
  formula for, 7
  and evolution, 111, 113
extinction, risk analysis, 106–107

falsificationist account of science,
    27–29
Fenchel, Tom, 17
Fenchel allometry, 17, 19–20, 21
Feynman, Richard, 121
forces, 101. *See also* ecological forces
friction, as analogy for limiting
    factors in population growth,
    98–99

Galilean relativity principle, 95
  as analogy for ecological invariance, 96–97
Galilean transformations, 96
Galileo, 4–5, 36–37, 41
  inertial view of, 4–6, 26, 36–39
Gause, Georgyi Frantsevich, 68
Gause loop, 67–69
generation-time allometry, 16,
    19–20
Gosse's creation theory, 121–122

homeotherms, allometries of, 13, 17
humans, and inertial growth, 53, 56
Hutchinson's evolution metaphor,
    112
hydra, 39–42, 43

inertia, 4–6, 36–39, 83, 98, 116. *See
    also* inertial population growth
inertial population growth, 6–8,
    9–10, 50–57, 83–84. *See also*
    cyclic population growth
  consequences for population
    management, 108–110, 114–
    116
  effects of artificial mortality on,
    107–109
  equation for, 90–91, 94–95
  evolutionary implications of,
    111–113
  experimental evidence of, 62–63
  implications for conservation
    114–116
  implications for risk analysis,
    113–114
  implicit model for, 83–89
  and maternal effect, 10, 49–50,
    50–52, 102, 130
  practical consequences of, 104–
    116
  and predator–prey interactions,
    57–59, 64–82
  simplicity of theory of, 117,
    125–129
instantism, 70–74
invariance, Malthusian, 65, 90–94

Kepler, Johannes, 21
Kepler's laws of planetary motion,
    21–24, 25, 26
Kleiber, Max, 12
Kleiber allometry, 12–15, 20, 21, 31,
    46

Lagopus scoticus, 92
Lakatos, auxiliary hypothesis of, 28
larch budmoth, 88, 93
laws of ecology, 11–12, 26, 30–33,
    101–103. *See also* ecological
    allometries
laws of nature, 26–30
  exceptions to, 26–27
  as explanations, 29–30, 31
  and falsifiability, 27–29

laws of nature (*continued*)
  misconceptions about, 26–30
laws of physics, 11, 31–32. *See also*
  analogies, scientific
lemmings, 55, 60, 61, 79, 93
Lemmus lemmus, 93
limit myths, 27, 65–66
Lotka, Alfred James, 64
Lotka–Volterra model, 65, 78, 81
  cycle period of, 81
Luckinbill, L. S., 68
lynx–hare cycle, 55, 60–61, 68–69

MacArthur, Robert, 114
Malthus, Thomas, 7, 9
Malthusian growth, 7–8, 35–36,
  96–98, 129. *See also* exponential
  population growth
Malthusian invariance, 65, 90–94
maternal effect, 9–10, 49–63, 102,
  130
  as ecological force, 101–102
  and inertial population growth,
   10, 49–50, 50–52, 102, 130
  inverse, 49
  and predator–prey interactions,
   78–81
  model and equation for, 50–52
  and problems with population
   data, 83, 84
Maxwell, James Clerk, 8
metabolism, and population growth,
  34, 43, 44–46, 46–48, 97–
  98
metaphors, in science, 8–9, 57. *See
  also* analogies, scientific
Microtus agrestis, 92
Minkowski, 119
mixed strategy, 49
muskox, 92

natural laws. *See* laws of nature
Newton, inertial view of, 4–6
Newton's laws of motion, 4–6, 95,
  100–103
Newtonian gravitational theory, 25,
  28–29, 31–32, 124
Newtonian mechanics, as analogy
  for inertial population growth,
  100–103

Ohm's law of electricity, 11
overfitting of scientific models, 79
overfitting of scientific theories, 116,
  123–125
Ovibos moschatus, 92

parental effect, 49. *See also* maternal
  effect
pest species management, 110, 115
physics, as analogies for ecology,
  3–4, 8–9, 57. *See also* analogies,
  scientific
physics, differences from ecology, 32
Pimm, Stuart, 75
planetary motion
  as analogy for population growth,
   3–10, 57
  Kepler's laws of, 21–24, 25, 26
  Titius–Bode law of, 23–24, 24–25
Plato's cave metaphor, 118–119
poikilotherms, allometries of, 13, 17
Poincaré, Henri, 121
population abundance, model for,
  44–46
population equilibrium, 43–46,
  46–48, 107–109, 113
  and analogy of falling body on a
   spring, 46
population growth. *See also* cyclic
  population growth, exponential

population growth, inertial
population growth
accelerated, 35–36, 37–41, 41–43,
43–46
and analogies with planetary
orbits, 3–10, 57
and delayed density dependence,
51, 94, 108
and density dependence, 51, 61,
89, 94, 108
equations for, 70, 90–91
limiting factors of, 65, 89, 98–100
and metabolism, 34, 43, 44–46,
46–48, 97–98
and reliance on individual
energetics, 44–46, 97–98
and response lags, 10, 48, 125
as a second-order process, 41–42,
44–46, 90–92
population size, formula for, 7
predator–prey interactions, 64–82
and Calder allometry, 57–59
and coexistence, 67–69, 109, 112
donor-controlled models of,
75–77
and dual extinction, 67–69, 112
and Gause loop, 67–69, 112
with generalist predators, 79–81
model parameters of, 78–79
and moose, 72, 73
and population cycling, 57–59,
64–82
and prey refugia, 68–69
and prey-dependent predation, 66,
66–70, 72, 123, 130
and ratio-dependent predation,
65–70, 72, 75–78, 97–98, 102
with specialist predators, 79–81
and wolves, 72, 73
prey refugia, 68–69

prey-dependent predation, 66,
66–70, 72, 123, 130

Quine, W. V., 27, 28

ratio-dependent predation, 65–66,
66–70, 72, 75–78, 97–98, 102
donor-controlled models of,
75–77
and equation for prey mortality,
75–76
red deer, 92
reductionism, 32
resilience, 109
resonance avoidance, 111–112
Reynolds number, 98
risk analysis of decline or extinction,
106–107
ruffed grouse, 92

scientific aesthetics, 30, 100, 120–
123, 124
scientific method, 120–123
scientific theories
and facts, 122–123
increasing dimensionality of, 117,
118–119
overfitting of, 116, 123–125
simplicity of, 117, 121, 123
shrews, and Kleiber allometry, 12
Slobodkin, Lawrence, 39, 42
and hydra experiment, 39–42, 46
starvation, effects on population
growth, 35–36, 39–42, 43, 87

theoretical ecology/ecologists,
104–106
and relations with applied
ecologists, 105–106
trophic chains/pyramid, 67, 112, 127

unicellular organisms, allometries of, 13, 17

Veilleux, B. G., 68
voles, 53, 60, 61, 79, 92, 93

Volterra, Vito, 64. *See also* Lotka–Volterra model

Zeiraphera diniana, 88, 93

DISCARDED
CONCORDIA UNIV. LIBRARY